Sedimentary Rocks
in the Field
A Color Guide

Dorrik A.V. Stow Ph.D

Institute of Petroleum Engineering
Heriot-Watt University
Edinburgh, Scotland, UK

D0061084

ELSEVIER

ACADEMIC PRESS
An imprint of Elsevier
books.elsevier.com

For Jay, Lani, and Kiah

Sixth impression 2012
Fifth impression 2011
Fourth impression 2010
Third impression 2009
Second impression 2006

First published in the United States of America in 2005
by **Academic Press, an imprint of Elsevier Inc.,**
225 Wyman Street, Waltham, MA 02451, USA
525 B Street, Suite 1800, San Diego, CA 92101-4495, USA

www.store.elsevier.com

ISBN: 978-0-12-369451-5

Commissioning editor: Jill Northcott
Project manager: Ayala Kingsley
Copy-editors: Kathryn Rhodes, Ayala Kingsley
Designer: Ayala Kingsley
Colour reproduction by Tenon & Polert Colour Scanning Ltd, Hong Kong
Printed by New Era Printing Company Ltd, Hong Kong

FOREWORD

SEDIMENTARY ROCKS IN THE FIELD: A COLOUR GUIDE is a much needed addition to the literature of sedimentology. It offers much more than its title reveals. Eleven of the fifteen chapters define the main types of sedimentary rock and illustrate them in all their beauty and variety with numerous color photographs. The remaining chapters – also lavishly illustrated – provide introductions to field techniques, principal characteristics of sedimentary rocks, and interpretations of depositional environments, all of which enrich the book tremendously. The provision of stratigraphic time scales, mapping symbols, grain-size comparator chart and sediment description checklist, together with Wulff stereonet and Lambert equal-area projection templates, further adds to its usefulness in and out of the field.

The combination of all these elements in a handy format will enable the geologist to more readily identify and understand the type of rock he or she is dealing with, making life in the field a lot easier! This publication will be of great help both to the student and to the professional who has not been in the field for some time, while amateurs, whose background on sedimentary rocks may be incomplete, will welcome it too. In addition to its value for geology students, professionals, and amateur enthusiasts, the book will also be of interest to anyone – from weekend walkers to soldiers on maneuver – who takes a moment to study the terrain.

Professor Stow has traveled extensively, both on and off-shore and that experience lends maturity to this publication. There are examples of rock outcrops from all over the world and the author provides different scales, such as micron, millimeter, and phi values for grain-size, in order to make this work internationally useful. Finally, the publisher and his team did a great job on the lay-out and production.

I strongly recommend this publication to any student, professional, amateur, or outdoor person interested in learning more about sedimentary rocks in the field.

Arnold H. Bouma Ph.D
Department of Geology and Geophysics, Louisiana State University

CONTENTS

PREFACE

THE WORLD OF sediments and sedimentary rocks is exciting and dynamic. It is fundamental to our understanding of the whole Earth System and of the wide range of environments that characterize its surface. It also provides the key to a plethora of natural resources – industrial, chemical, metallic, water, and energy resources – that shape the way we live.

Ideas and concepts in sedimentology are fast changing, but fundamental fieldwork and data collection remain at its heart. In the first instance, it is an observational science, closely followed by laboratory, experimental, and theoretical work. The primary skill lies in knowing how and what to observe and record in the field, and then how best to interpret these data. For me, this has always been a distinctly visual process. The unique aspect of this guide, therefore, is in the wide range of graphic material that draws together the very latest ideas and interpretations (over 50 line drawings), coupled with over 425 photographs (from 30 different countries) of the principal types of sedimentary rocks and their characteristic features. It is intended for ease of field use by students, professionals, and amateurs alike.

All the field photographs illustrated have been carefully selected from my own collection, except where otherwise acknowledged. All figures have been redrawn and many specially compiled from the latest research knowledge, always with a view to providing the best aid to recognition, classification, and interpretation in the field. The key emphasis is to help with field observation and recognition of the main features of sedimentary rocks. Some pointers are given towards their preliminary interpretation, but further endeavour in this area must remain the province of the broader sedimentological literature, and will depend on the nature of the work in progress. Many different disciplines and sub-disciplines of geology and oceanography, as well as sedimentology, require a field understanding of sediments and sedimentary rocks. They include: geophysics and geochemistry, paleontology and Quaternary geology, physical geography and soil science, archeology and environmental science. Above all, and for all, this is a book to take into the field and use!

ACKNOWLEDGMENTS

THERE ARE very many people to acknowledge in compiling a book of this sort. Most importantly, my thanks to Claire for the original idea, one wet and rainy day in the Lake District, and for her continued support throughout, and to Michael Manson for his enthusiasm and enduring patience through a long gestation period. Thanks also to other members of the Manson Publishing team and to Ayala Kingsley for expert design and layout. At my own institute, Southampton Oceanography Centre (now the National Oceanography Centre), I am much indebted to Frances Bradbury for typing the first manuscript, Kate Davies for such expert drafting of the many complex figures, and Barry Marsh for hand specimen photography. Particular thanks are also due to the principal reviewer of an earlier version of the manuscript, Tony Adams, for his hard work and helpful comments, and to Ian West and Vicky Catterall for subsequent review and proof reading.

Great appreciation goes to numerous colleagues the world over, who have introduced me to such interesting rocks and have shared time, ideas, and friendship in the field, as well as to countless students who have so aptly demonstrated the need for such a book. Paul Potter, my long-time American colleague who roams the western hemisphere, deserves special mention – his numerous, well illustrated books on sedimentology have inspired, and a number of his excellent field photos used here have helped plug gaps in my own collection. Hakan Kahraman has been particularly helpful with the chapter on coals, Ian West on evaporites, and Bob Foster with the chapter on ironstones. Special thanks are also due to Arnold Bouma for inspiration and insight in the field, for reading and commenting at proof stage, and for kindly providing a generous Foreword. Institutional and financial support has come from varied sources, in particular the Southampton Oceanography Centre at Southampton University, the Royal Society, BP, the British Council, and the Natural Environment Research Council (UK).

OVERVIEW

Planning the work. Mixed volcanic and volcaniclastic succession, Lake District, UK.

About this book

THIS BOOK is intended as a guide to the recognition and description of sedimentary rocks in the field. The emphasis is firmly on illustrating the principal types of sedimentary rock and their specific characteristics through a series of colour photographs and summary figures. Lists and tables of key points and features keep the text succinct. It is designed for ease of use in the field, with a careful cross-referencing and index system, and with some of the key tools of the trade printed on the cover, or as figures at the end of the book (see pages 314–320). The first steps towards interpretation are included, together with a guide to more detailed texts for further work.

This introductory chapter briefly defines the main types of sedimentary rock and their recognition, followed by a section highlighting the economic study of sediments. Chapter 2 describes the main field techniques including safety, equipment, field notebook records, making basic measurements, field sketches and logs, paleocurrent analysis, and stratigraphic procedure. Chapter 3 provides a comprehensive summary of the principal characteristics of sedimentary rocks including bedding, sedimentary structures, sediment texture and fabric, composition and colour, as well as how to recognize these features, and what to observe and measure. This chapter is fully illustrated with field photographs. Chapters 4–14 document each of the main rock types in turn and further illustrate these with a wide range of field photographs. Chapter 15 gives a brief summary of how to interpret sediment facies and their features in terms of depositional environments. Finally, there are also useful appendices featuring time scales, mapping symbols, a grain-size comparator chart, and Wulff stereonet and Lambert equal-area projection templates.

The book has been written primarily for university students in earth and environmental sciences, as well as for professionals in a range of related disciplines. However, it is also readily accessible to senior high school students and to the amateur geologist. There are over 50 line drawings and over 425 colour photographs of rocks in the field or from core sections. These are drawn from every geological period and some 30 different countries around the world.

Classification of sedimentary rocks

SEDIMENTARY ROCKS are formed from the accumulation of particulate material through physical, chemical and biological processes at the surface of the Earth. These processes operate in a whole range of natural environments, most of which exist today, so that sediments (unlithified) and sedimentary rocks (lithified) currently cover some 70% of the Earth's surface. They also make up a significant proportion of the geological record.

Sedimentary rocks can be distinguished from igneous or metamorphic rocks on the basis of one or more key characteristics including:

- Their organization into subparallel strata or beds.
- Their composition of discrete particulate material – detrital mineral grains, biogenic debris, and transported clasts – together with a binding cement.
- The presence of sedimentary structures and fossils.

They are generally subdivided on the basis of origin, as reflected by composition, into four principal groups defined as follows:

- Terrigenous sediments (also siliciclastic or clastic) are composed mainly of detrital grains derived from the weathering and erosion of any pre-existing rock – they include conglomerates, sandstones, mudrocks and some palcosols.
- Biogenic sediments (also bioclastic or organic) are derived from the skeletal remains and soft organic material from pre-existing organisms, and of material that has been biosynthesized – they include carbonates (limestones and dolomites), some cherts, phosphorites and coal. Whereas most limestones and dolomites were originally derived from biogenic material, some limestones are chemical precipitates (e.g. oolites and travertine), and most dolomites are formed by post-depositional chemical alteration of the original carbonate. Classifications are never perfect!
- Chemogenic sediments (also chemical or authigenic) are those formed from the direct precipitation of crystalline particulates from concentrated saline solutions – they include evaporites, ironstones and other metalliferous sediments, some carbonates (see above) and duricrusts.
- Volcaniclastic sediments (also volcanigenic or pyroclastic) are composed mainly of grains and clasts derived from contemporaneous volcanic activity – they include autoclastites, pyroclastites, hyaloclastites and epiclastites, (also ash, tuff and agglomerate).

Within each major group or subgroup, sediments can be further subdivided into a variable number of sediment facies (Chapter 3). These are based on specific characteristics – principally sediment structures, textures, and composition. They are the building blocks for further interpretation (Chapter 15).

Economic studies

SEDIMENTS and sedimentary rocks have a very wide range of uses in industry and society in the contemporary world. Indeed, without the exploitation of sedimentary materials, the world would be a very different place. Some of the principal industries, products, and their sedimentary sources are listed in Table 1.1.

Field sedimentology, therefore, is very concerned with its economic application. Energy and water resources, metals and industrial minerals, and the raw materials for chemical and domestic products are all to be found in sediments. For any particular product or use, the sedimentary source material must be located (exploration, mapping) and then assessed in terms of its abundance, distribution, quality and extractability. These goals are achieved through careful fieldwork, the development of appropriate sedimentary environment models, and subsequent testing in the laboratory. It is hoped that this book will go some way towards helping with applied sedimentology in the field as well as with the laboratory study of core material.

Table 1.1 Economic use of sedimentary materials

Industry	Product/Use	Sedimentary source
Construction industry (road, buildings, transport network, etc)	Concrete, mortar	Limestone, gravel, sand
	Building stones, bricks	Various, clays
	Roofing tiles, slates	Clays, sand, lime, coal, slate
	Insulation	Various
	Glass	Silica sands
	Plaster, plaster board	Gypsum, anhydrite
	Road metal, bitumen	Various, hydrocarbons
Domestic Products	Plastics, paint	Hydrocarbons, clays
	Foodstuffs	Various
Chemical industry	Petrochemical, plastics, etc	Hydrocarbon products
	Pharmaceuticals	Various
	Agricultural products	Phosphorites
	Other, lubricants, fillers	Evaporites, ironstones, siliceous sediments
Metal industry	Iron and steel products	Ironstones, coal,
	Other metals	Placer deposits
Energy industry	Oil and gas	Sedimentary source and reservoir rocks
	Coal and peat	Coal, peat
	Nuclear industry	Uranium in sediments
	Geothermal power	Hot water aquifers
Water industry and waste disposal	Drinking/industrial water	Sediment aquifers
	Waste disposal	Various sediments

1.1 Photograph and sketch section of a Plio–Pleistocene fan-delta succession, southern Cyprus. Sand-body geometry is fundamental to an assessment of both hydrocarbon reservoir and freshwater aquifer properties. Outcrop analogues are sought by oil companies in order to improve understanding and prediction of sand-body architecture in subsurface reservoirs. In this case, lenticular and elongate sand bodies are formed by rapid progradation of channel sands across a fan-delta top. This is fed by powerful, short-headed rivers draining the interior of Cyprus. Each major sand unit is capped by a paleosol horizon, which represents a period of inactivity during a more arid climatic phase

FIELD TECHNIQUES

Nose to the rocks. Detailed field observation of discontinuity, Osmington Mills, S. England.

Safety in the field

GEOLOGICAL fieldwork carries its own special hazards. In the first instance, each individual is responsible for his or her own safety as well as being mindful of others' welfare. Everyone should also be environmentally aware at all times. Be prepared for all conditions and eventualities.

- Be aware of specific hazards in the area of work, including weather forecasts, tides, sea conditions, dangerous animals/insects and so on.
- Carry the appropriate emergency equipment, including first aid kit, whistle, flashlight, matches, space blanket, water, provisions and glucose tablets, and wear suitable clothing and footwear.

- Learn some first aid - take a course, carry a leaflet, learn the signs and treatment for exposure, take care with drinking water.
- Advise a colleague, friend, base camp, hotel, coastguard or other appropriate person of your planned itinerary each day.
- Take a mobile phone with you whenever possible – the international emergency number is 112; learn the local number.

Specific precautions
- Special care with cliffs, quarry walls or other steep faces.
- Special care with loose rocks on steep slopes, for yourself and those nearby - wear a safety helmet where necessary.
- Be aware of potential rock falls, land-slides, mudflows, sinking mud/sand, flash floods, earthquakes and so on.

- Keep a keen eye out for rogue waves and slippery rocks along the shore.
- Avoid running down very steep slopes or climbing rock faces/cliffs unless suitably trained and with a companion.
- Avoid old mine workings and caves, unless suitably trained and with appropriate authorization.
- Take great care with hammering/collecting – wear safety goggles and keep away from others. Avoid hammering chert.
- Be mindful of any old equipment, litter (e.g. cans, glass), explosives (in quarries), traffic, and heritage/preservation orders.

International distress signal:
- 6 whistle blasts/ flashes of light/shouts/ waves of bright-coloured cloth, repeated at one minute intervals as necessary.
- 3 whistle blasts to acknowledge a distress signal.

Field equipment

IT IS ESSENTIAL to be well equipped for fieldwork. The following items are the most important:

- Backpack (strong, lightweight, with many pockets).
- Geological hammer (around 1kg is generally adequate).
- Compass clinometer (remember to set for magnetic declination in the study area).
- Hand lens (x10 magnification is common – take a spare).
- Penknife (with corkscrew and bottle opener for relaxation and discussion!).
- Tape measure (a long reel-measure if extensive logging is planned).
- Acid bottle (sufficient for duration of work).
- Field notebook (hardback, weather resistant, large enough for sketches, logs and a neat, ordered record – minimum size 180 x 220mm recommended here).

- Grain-size and sorting comparators, stereonet and overlay paper.
- Pens, pencils, ruler and coloured crayons for mapping, drawing pin for stereonet.
- Sampling bags and marker pens.
- Base maps and clip-board (if mapping).
- Graphic log-sheets and folder (if much logging is planned).
- Topographic maps, geological maps and relevant literature.
- Camera, smile and cheerful mood.
- This book!

For safety and comfort, take or wear:
- Appropriate clothing (bear in mind sudden changes in weather conditions).
- Appropriate footwear and goggles.
- Small first aid kit, whistle and flashlight.
- Emergency rations and adequate water.
- Sun lotion.

This may seem like a long list for a single excursion but, in fact, it's the minimum for any serious work. Additional items will be required for special conditions. These include: a trowel and spade for work with modern or unconsolidated sediments; one or more chisels for any serious collecting; and chemical stains (e.g. Alizarin Red Stain) for work with carbonates. A pair of binoculars can also be very useful for observing distant exposures as a preliminary to further examination, and for inaccessible crags or cliffs.

Portable GPS (Global Positioning System) units are now readily accessible and relatively inexpensive. They can provide a useful addition to field equipment, but are no substitute for constant field awareness in relation to the base map. A range of other electronic measuring devices now exist – e.g. field-operated spectrometers for geochemical determination, hand-held gamma-ray meters for source rock evaluation.

Notebook or laptop computers can record data in the field, and may incorporate base maps, aerial photographs, a GPS device, and proformas for specific data types and logging.

14

General approach and field notebook

THE AIMS of the study, time available, nature of outcrop and weather conditions will determine the detail afforded at any one locality. However, whether undertaking regional reconnaissance or detailed section logging and collection, the general approach is the same: be systematic, scientifically objective, highly observant, ask questions and test theories, and record all information/ideas in a field notebook or hand-held computer.

The field notebook is one of the most important items for any student or professional geologist alike. It is the original scientific record of your observations. It is likely to be referred to again and again throughout your career, so keep it neat, systematic, and thorough. Add incidental notes of interest (people, nature, weather, wine...) as *aides memoires* for future reference, and ensure there is a current name/address in the front for easy return if lost.

Many geologists use standard surveyor's notebooks (125 x 200mm page size), but this is too small for good sedimentological work. I recommend a minimum page size of 180 x 220mm and preferably 200 x 250mm. These sizes are more appropriate for field sketches and logs. Notebooks should be hardback with squared (or ruled) pages and, ideally, a water-resistant cover. Purchase in advance and always take spares.

USEFUL TIP
Keep a backpack packed with field gear and ready to go at all times. Be prepared for all conditions. Make yourself comfortable to work – e.g. sit on a rock or ledge to sketch sections; retire to a bar or café when it rains. Detailed fieldwork can be arduous, but is extremely rewarding when good observations are made and good data have been collected.

2.1 Sample field notebook entry. This illustrates a fairly full entry for a single locality, showing 'good practice' systematic approach for recording data.

The principal information to record for each main locality is as follows (Table 2.1, see also *Fig. 2.1*). Use standard abbreviations (see page 317), and devise your own where necessary. Use the correct geological and sedimentological terminology, as presented in this book – learn the language.

Basic measurements and data records

MEASURING and recording data in the field for completing notebook entries is mostly a matter of common sense and will vary according to need and conditions. Summary guidelines are given below.

Strike and dip, trend and plunge
(*Fig. 2.2*)

- *Planar surfaces* – e.g. bedding planes, foreset cross-beds, fault planes – are recorded by strike and dip; a technique for measuring these parameters using a compass clinometer is shown in *Fig. 2.2*.
- *Linear features* – e.g. bedding plane lineations, axial trend of flutes, preferred orientation of elongate clasts – are recorded by trend and plunge (or pitch where easier to measure).
- The *quadrant* method for recording strike and dip is to note the bearing of the strike, an oblique stroke, and then the amount of dip and the quadrant it points to – e.g. 042°/30°SE. This gives a choice of two strike readings at 180° to each other (i.e. 042° and 222° in the case above), of which I routinely use the lower, although either is correct. This is the method I recommend as I believe it less confusing than other techniques (below) and keeps the observer always thinking about orientation in the field.

Date: 3.8.95 (Cyprus)

LOC: Pissouri Village, near water tower GR 7245 3671
roadside section just N of village centre on road up to water tower
– typical sst facies on which village stands, and probably makes up
caprock of other highs in Pissouri Basin

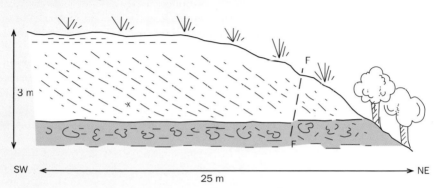

3 m

SW ←————————————————————→ NE

25 m

Strat: probable Quaternary section - near top of basin fill
Bedding: v. gently dipping interbedded grey-yellow sst and red mdst
Sst. bed ~ 2.5m thick (⟵), overlain by thin-bedded (≡) sst,
underlain by nodular mdst ~ 0.7m thick

025 / 10° NW
032 / 8° NW

Structures: main sst with lg-scale cross bedding (set 2–2.5m),
intensely burrowed esp. in lower part of
foresets 010 / 26° SE
008 / 27° SE
015 / 24° SE

approx
vertical
burrows

red mdst – structureless + irregular CO_3 concretions

Textures: sst–med-g, mod–poor sorting; mud–silty

Composition: sst–lithics (volc, cryst. ig.) ~ 60%
CO_3 (mainly micrites) ~ 30%
Qtz ~ 5%, matrix ~ 5%
mdst – v. red color + irreg. CO_3

Other: calcrete covering upper surface of exposure
minor normal F ~ 170 / 85° SW, minimal displacement ~ 3cm

Comments: Gilbert-type x-beds – possible braid △ channels
feeding shallow-marine environ; red mdstn ≡ typical thick paleosol
horizon (? pluvial / interpluvial periods)

- For the *right hand rule* method, use your right hand with forefinger and thumb at right angles. Align the forefinger along strike and thumb down dip. This gives a unique strike direction, with no need to note quadrant for the dip. The above example would be written 222°/30°. The dip and dip direction method does not record strike at all, so that the above example would be written as 30°/132°. However, strike is, of course, a key reading for structural information.

Sediment facies

Sediment facies (or lithofacies) are the basic building blocks of field sedimentology. Simply expressed, they are the multifarious sediment types that exist, each distinguished from other facies by their distinctive attributes (see page 28, and Chapter 3). The first task, therefore, is to establish sediment facies.

- Determine by close inspection which sediment facies are present at a particular locality. Remember, there is no absolute facies type – the number and detail of facies selected will depend on the scope of

Table 2.1	Standard sedimentological observations to make for each locality
Locality	Record location details (map number, grid reference, name, etc).
	Note any particular reasons for selection of this locality (e.g. type-section, testing theory).
General relationships [± field sketch]	Bedding – way-up, dip and strike, stratigraphic relationships.
	Structural aspects – folds, faults, joints, cleavage, unconformities, intrusions, veins and so on (measure spacing, sense, orientation, etc).
	Weathering, vegetation – may reflect or obscure lithologies.
	Topography – may reflect lithologies, structures.
Lithofacies and units	Note principal lithofacies present (sedimentary and other rock types).
	Note any larger facies associations, sequences, cycles, mapping units and so on.
	Note state of inclination and/or metamorphic grade.
Detailed observations [± specific sketches or logs]	Bedding – geometry, thickness, etc.
	Sedimentary structures (including paleocurrent data).
	Sediment texture and fabric.
	Sediment composition and colour, including biogenic material.
	Any other information.
Questions and theories	Note down any questions, problems, ideas, plans for sampling and lab work. Much of the above information as well as additional data can be recorded by use of field sketches, logs, photographs, sample collecting, and so on (see the following four sections).

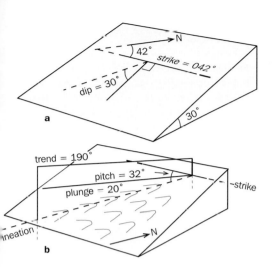

2.2 Field measurement of (a) dip and strike of planar surface (e.g. bedding plane), where strike = bearing of horizontal line on the surface of the plane, and dip = angle from the horizontal of maximum slope of the plane (ie at right angles to strike); and (b) trend, plunge and pitch of lineation on a planar surface, where trend = bearing of an imaginary vertical plane passing through the lineation, plunge = inclination of the lineation in that plane, and pitch = acute angle the lineation makes with the strike of the surface in which it occurs.

To measure strike:
1) *use notebook/map case to smooth surface if necessary; place edge of compass on this surface, hold horizontally, align parallel to strike and read bearing;*
   OR
2) *mark strike line on surface, stand over it or look along it, align compass parallel to strike and read bearing.*

To measure dip:
1) *place clinometer on rock surface (or notebook etc) at right angles to strike line and measure angle of dip;*
   OR
2) *estimate angle of dip using clinometer and standing end-on to the exposure.*

Record measurements by Quadrant Method: bearing of strike/amount of dip and compass quadrant it points towards, ie 042°/30° SE. Measure trend and plunge directly, as for strike and dip, or measure pitch together with the strike and dip of the planar surface in which it occurs.

the study. For example, for broad reconnaissance mapping simple facies divisions are adequate (e.g. conglomerate, sandstone, mudstone), whereas for detailed sedimentological analysis a more sophisticated breakdown of each of these groups into specific types will be required (e.g. parallel-laminated sandstone, wave-ripple cross-laminated sandstone, etc).

- Careful logging of facies and facies transitions through an extended section is necessary for subsequent statistical analysis of vertical facies sequences.
- Note also the lateral variation of facies characteristics within the vicinity.

Bed thickness

- For *outcrop/lithofacies* characterization – measure maximum and minimum bed thickness, make visual estimate of modal thickness, and record any thickness variation in individual beds.
- For *logged sections* – measure bed thickness in line of section and note any variations away from section (eg wedge-shape, lenticularity, etc.) Remember, there is a standard terminology for bed thickness – see *Fig 3.1*.
- For *statistical analysis* of bed thickness variation (for cycle/sequence analysis back in lab) – measure >100 beds in sequence (50 minimum for small-scale cycles).

Sedimentary structures.

- For *all structures* – look and record carefully the variety and intercalation of all sedimentary structures; scout around to find the best outcrop for revealing structures, cleaning up certain exposures as necessary.
- For *paleocurrent data* – collect as many different measurements as possible (see page 23); single measurements of cross-lamination directions are notoriously misleading.

- For *structural sequences* (in turbidites) – record the relevant divisions present for each turbidite bed on graphic log, i.e. TABC (for Bouma divisions), T01238 (for Stow divisions) and so on.
- For apparently *structureless beds* – clean and examine the surface very carefully; consider taking samples back to the lab for more detailed study of polished slabs.
- For *biogenic structures* – look carefully for any evidence of trace fossils and bio-turbation, as these can be as important as primary sedimentary structures (see page 86, Chapter 3).

Sediment texture and fabric

- For *grain size/sorting* – use comparator charts and hand lens where possible, or ruler/tape measure for larger clasts; measure 200 clasts in a 0.5 x 0.5m square (quadrat), estimate modal size and note the maximum clast size present.
- For *grain morphology* – use comparator charts and hand lens.
- For *sediment fabric* – orientation of long grains/clasts/fossils can be made efficiently using a transparent acetate overlay and marker pens (20 readings if very uniform, 50+ for greater statistical accuracy).

Sediment composition

- Examine several different hand specimens or parts of a particular bed in order to establish internal variability and to record the presence of minor accessory components – a hand lens and nose-to-the-rocks are essential for most sediment types.
- For *coarse-grained fabrics* – rapid counts can be made of 50+ clasts in a 0.5 x 0.5m quadrat.
- For *carbonates* – a drop of 10% HCl on a fresh surface shows vigorous efferves-cence with calcite, and slow effervescence on a freshly powdered surface with dolomite; Alizarin Red S mixture stains calcite red but leaves dolomite (and quartz) unstained.

Field sketches and logs

FIELD SKETCHES and logs are as important or more important than many words of description. They allow a much quicker and more accurate record of features and relationships and are better for later recall. You do not need to be an artist – they should be schematic, but accurate in terms of relationships and relative scale; they should be simple, focusing on sedimentary and structural aspects rather than vegetation, weathering features, and so on; and they should be well annotated. Always remember to add scales, way-up and compass orientation. Use standard abbreviations (see page 317), and devise your own additional ones as needed.

Field sketches (*Fig. 2.3*) represent lateral relationships and bed geometries of an exposure, the general stratigraphic section, relationship with any igneous/intrusive bodies, facies associations or mapping units, and large-scale sequences/cyclicity.

Sketch logs (*Fig. 2.4*) represent the vertical succession through part of an exposure, and are useful for showing stratigraphic order, vertical sequences and principal lithofacies.

Graphic logs (*Fig. 2.5*) are accurately meas-ured sedimentary logs through a vertical suc-cession. They provide an excellent way of sys-tematically recording detailed sedimentary data in the field. Graphic logs can be drawn either in your field notebook or on prepared logging sheets (with clipboard) – these have the advantage of providing a standard format and ready checklist.

2.3 **Sample field sketch as field notebook entry. Good practice with full details. Note the use of partial ornamentation at left of section to avoid obscuring key structural features as sketched.**

Hill Head near Lee-on-the Solent / southern England
low cliff section approx 1km NW of Titchfield Haven, Hill Head

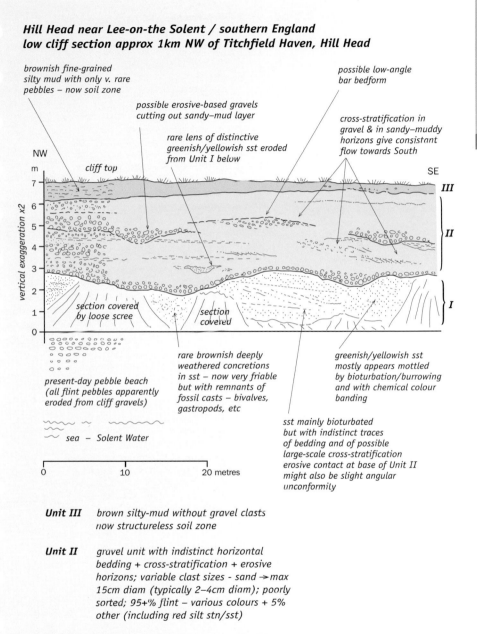

brownish fine-grained
silty mud with only v. rare
pebbles – now soil zone

possible erosive-based gravels
cutting out sandy–mud layer

rare lens of distinctive
greenish/yellowish sst eroded
from Unit I below

possible low-angle
bar bedform

cross-stratification in
gravel & in sandy–muddy
horizons give consistant
flow towards South

NW

cliff top

m

7

6

5

4

3

2

1

0

SE

III

II

I

vertical exageration x2

section covered
by loose scree

section
covered

present-day pebble beach
(all flint pebbles apparently
eroded from cliff gravels)

rare brownish deeply
weathered concretions
in sst – now very friable
but with remnants of
fossil casts – bivalves,
gastropods, etc

greenish/yellowish sst
mostly appears mottled
by bioturbation/burrowing
and with chemical colour
banding

sst mainly bioturbated
but with indistinct traces
of bedding and of possible
large-scale cross-stratification
erosive contact at base of Unit II
might also be slight angular
unconformily

sea – Solent Water

0   10   20 metres

Unit III brown silty-mud without gravel clasts
now structureless soil zone

Unit II gravel unit with indistinct horizontal
bedding + cross-stratification + erosive
horizons; variable clast sizes - sand →max
15cm diam (typically 2–4cm diam); poorly
sorted; 95+% flint – various colours + 5%
other (including red silt stn/sst)

Unit I greenish/yellowish glauconite sandstone

Hill Head near Lee-on-the-Solent, cliff section 1km NW Titchfield Haven

UNIT III brown silty-mud *(now soil zone)*
rootlet traces at top, rare flint pebbles, appears
structureless

Unit boundary sharp and distinct in most parts

UNIT II greyish flint gravels
structures: poorly defined, thick, irregular
bedding, crude horizontal and large-scale
cross-stratification, rare ripple cross-strat.
in finer-grained intervals, cross-cutting/internal
erosion features common
texture: grain-size from sand →15cm diam
gravel, typically 2–5cm, poorly-sorted, irregular
lateral & vertical gz variation. Clasts generally
sub-angular to angular (rarely rounded),
clast-supported fabric
composition: >95% flint pebbles, mostly grey
and brown coloured, also black, red and whitish
(both Fe and Mn oxide coatings); <5% other
clasts, including reddish siltstone/sst + 'rip-up'
rafts of glauconite sst and shales; no shell debris
observed

Unit boundary sharp and erosive, broad shallow channel
forms locally; possible slight angular unconformity
with UNIT I

UNIT I greenish-yellow glauconite sands
structures: v. indistinct parallel bedding /
lamination apparent in parts, also chemical
'liesegang' colour banding evident – partly
following original stratification; rare small-
scale cross-lamination, possible lg-scale x-
strat; mostly structureless → mottled, probable
intense bioturbation, some clear burrows
texture: med-grained sst. uniform, mod-sorting,
friable
composition: glauconite-rich quartz sand, now
with much decomposition to iron oxides etc.,
rare fossils in weathered concretions

highly weathered
concretion with
mollusc fossils
in crumbling
state of preservn.

pebble beach

metres

clay | silt | f m c sand | gravel → grain size

2.4 Sample sketch log as field notebook entry *(above)*. Good practice with full details. Note that increasing grain size is shown schematically along the x-axis in all such logs.

2.5 Sample graphic measured log on separate logging sheet *(right)*. Log sheets can be used for both fieldwork and core logging.

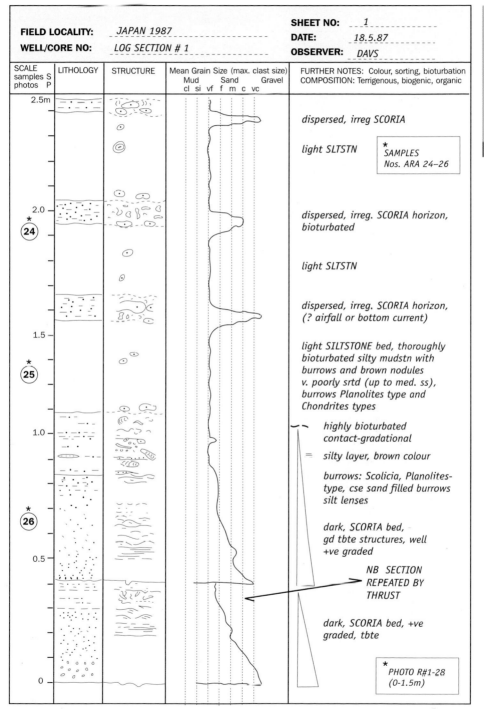

FIELD LOCALITY: JAPAN 1987

WELL/CORE NO: LOG SECTION # 1

SHEET NO: 1

DATE: 18.5.87

OBSERVER: DAVS

SCALE samples S photos P	LITHOLOGY	STRUCTURE	Mean Grain Size (max. clast size) Mud / Sand / Gravel cl si vf f m c vc	FURTHER NOTES: Colour, sorting, bioturbation COMPOSITION: Terrigenous, biogenic, organic

2.5m

dispersed, irreg SCORIA

light SLTSTN

*
SAMPLES
Nos. ARA 24–26

2.0
*
(24)

dispersed, irreg. SCORIA horizon, bioturbated

light SLTSTN

dispersed, irreg. SCORIA horizon, (? airfall or bottom current)

1.5
*
(25)

light SILTSTONE bed, thoroughly bioturbated silty mudstn with burrows and brown nodules v. poorly srtd (up to med. ss), burrows Planolites type and Chondrites types

1.0

highly bioturbated contact-gradational

silty layer, brown colour

burrows: Scolicia, Planolites-type, cse sand filled burrows silt lenses

*
(26)

dark, SCORIA bed, gd tbte structures, well +ve graded

0.5

NB SECTION REPEATED BY THRUST

dark, SCORIA bed, +ve graded, tbte

*
PHOTO R#1-28
(0-1.5m)

0

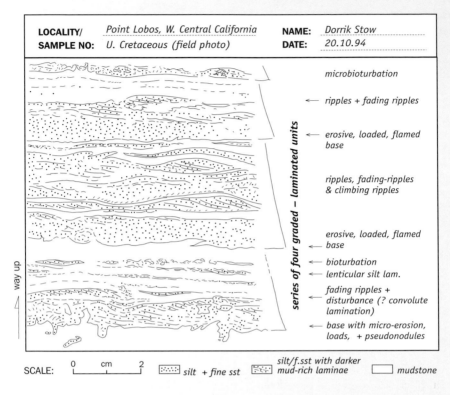

LOCALITY/	Point Lobos, W. Central California	NAME:	Dorrik Stow
SAMPLE NO:	U. Cretaceous (field photo)	DATE:	20.10.94

way up

series of four graded – laminated units

← microbioturbation

← ripples + fading ripples

← erosive, loaded, flamed base

ripples, fading-ripples & climbing ripples

← erosive, loaded, flamed base

← bioturbation

← lenticular silt lam.

← fading ripples + disturbance (? convolute lamination)

← base with micro-erosion, loads, + pseudonodules

SCALE: 0 cm 2 ⌈·····⌉ silt + fine sst ⌈░░░⌉ silt/f.sst with darker mud-rich laminae ▭ mudstone

Brief Description: (structure, texture, composition, other)
Interbedded fine-grained terrigenous sediments (silt/sandstone + mudstone)
Structures:
apparently series of 4 (or 5) graded–laminated units, with
thicker / coarser lamina (or v. thin bed) at base showing variable degrees of
erosive microscours, loads, flame structures and detached pseudonodules;
basal lamina is structureless – laminated and capped by fading-ripples /
probable convolute lamination also evident at this level; thicker zone of
ripples, fading-ripples climbing ripples in middle unit; units grade upwards
to thinner laminae, lenticular laminae (with micro cross-lamination – long
wavelength / low amplitude) and a micro-bioturbated zone near top, some of
larger, 'disturbed' zones may be burrows
Textures:
silt–fine sst interbedded with silty mudstone in graded laminated units
Composition:
terrigenous; sands quartz-rich, some feldspar

Interpretation: (Classification, origin, etc.)
Series of thin-bedded, fine-grained turbidites displaying Bouma CDE sequences
or Stow T0–T5 sequences. Turbidity currents from left to right in sketch

Key points to note for all types of log, either those made in the field or from core description, are as follows:

- Use appropriate scale – 1:5 or 1:10 for detailed work, 1:100 or 1:200 for reconnaissance.
- Log from base of succession upwards (the reverse convention is used for logging cores); no need to log in a continuous vertical line – offset to avoid hazardous or non-exposed sections.
- Bed thickness and boundaries – log to scale and note nature of contacts, group thin beds/laminae as necessary.
- Lithology – use separate column and standard symbols (page 317).
- Sedimentary structures – use separate column, sketch actual structures where convenient/interesting and/or use standard symbols for speed and ease (page 317); add paleocurrent directions.
- Sediment texture – use horizontal scale for recording grain size (coarser to right, finer to left), may be modified for carbonates and evaporites if required.
- Other features – composition, colour, samples, photos, detailed sketches, questions/comments.

Detailed field drawings (*Fig. 2.6*) are accurate representations of specific features (e.g. sedimentary structures, fossils, trace fossils, type facies and so on). They may help with subsequent interpretation, be used in reports or publications, and serve as a permanent record of your primary observations. Good photographs may replace drawings – but take care to record the location and purpose for each photograph in your notebook. This is particularly important when several hundred photos may be taken daily using a digital camera.

2.6 **Detailed field drawing as field notebook entry. Good practice for very detailed observation of sedimentary features in the field or in core section.**

Paleocurrent and paleoslope analysis

AN IMPORTANT data set that is most easily collected in the field is information regarding paleocurrent direction and the orientation of paleoslopes. In addition to current and wind directions, these data are important for the interpretation of facies, depositional environments and paleogeography.

Data types

Many different features in sedimentary rocks can yield paleocurrent or paleoslope information. Some record the actual direction of movement (the azimuth), whereas others record only the line of movement (the trend). The principal features are as follows – see appropriate figures for determination of current direction/trend and paleoslope orientation.

- *Erosional structures* (page 34, *Fig. 3.3*): most of these give current trend, flutes and chevrons give azimuth, large-scale erosional features are less easy to interpret without very good exposure.
- *Cross-lamination/bedding/stratification* (page 46, *Figs. 3.5–3.9*): these can yield good current azimuths, but for all scales of cross-lamination and bedding it is important to find 3D sections or bedding plane exposure to ascertain the maximum angle of dip of the lee-face; for trough-cross bedding this is essential – take great care with interpretation.
- *Contorted laminae and flame structures* (page 76, *Fig. 3.14*): these may show asymmetry and overturn in the direction of the paleocurrent.
- *Slump folds* (page 68, *Fig. 3.13*): these provide the best means of determining paleoslope (fold axes parallel to strike of paleoslope, fold overturn downslope), although the slope may be local rather than regional.

(a)

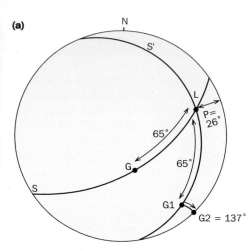

(b)

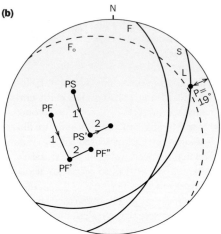

2.7 (a) A worked example of re-orientation of groove marks. Field data plotted as shown: bedding (S) 055/60°SE; intersection lineation (L) pitches 30°NE on bedding; groove marks (G) pitch 85°SW on bedding.

1. Rotate bedding (S) about the fold axis (L) to position where dip = plunge (P). G moves to G1 such that the angles L–G and L–G1 are both 65°.
2. Rotate the bedding (S') to horizontal. G1 moves along a small circle to give the original direction of lineation, G2 = 137°.

(b) A worked example of re-orientation of cross-lamination. Field data plotted as shown: bedding (S) 040/40°SE; foreset lamination (F) 010/48SE; intersection lineation pitches 30°NE on bedding. Plot the poles to the bedding (PS) and foreset lamination (PF), as well as the bedding (S) and foreset lamination (F) planes, and then plot intersection lineation (L).

1. Unfold by rotation about the fold axis (L) to position where dip = plunge (P). PS and PF move along great circles to PS' and PF' as S rotates about L. Position PS' is determined by L. The angle PF–PF' = PS–PS'.
2. Rotate the bedding to the horizontal. Return PS' to the centre along along small circle and move PF' the same angle along small circle to PF''. This is the pole of the original foreset orientation (F₀).
3. Read off the direction and amount of dip – i.e. 25° to 040.

From Tucker 1998

Continued from previous page...

- *Glacial striations* (page 36–37): where these are present on bedrock they show the direction of ice movement.
- *Sediment fabric* (page 113, *Fig. 3.31*): the preferred orientation of elongate clasts and fossils gives current trend, although pointed clasts (e.g. belemnites) will generally lie with the point upcurrent; an imbrication fabric yields current azimuth.
- Regional trends in many different parameters can be used to infer both paleocurrent and paleoslope information – these include trends in grain size, composition and bed thickness (all tending to show a decrease away from source), trends in facies and facies association, fossil and trace fossil assemblages and so on.

Data analysis

The most accurate and complete interpretations will be made by:

- Measuring a number of different data types (see above), as each may yield a different part of the picture.
- Taking measurements separately from different lithofacies and beds.
- Making as many readings as possible to achieve a statistically better result.

Attention must also be given to potential changes in orientation brought about by tectonic activity:

- Tilting of beds by more than about 10–15°. This can be relatively easily corrected for by means of a Wulff stereonet (page 319) following the procedure illustrated in *Fig. 2.7*.
- Rotated thrust slices or larger crustal slabs. This can only be corrected for where the nature and degree of rotation has been independently determined (e.g. by paleomagnetic measurements).
- More intense deformation of strata (involving folding, cleavage, metamorphism and so on) will also deform the paleocurrent structures, and so severely restrict the accuracy of any paleocurrent analysis.

Paleocurrent (and paleoslope) data are most conveniently plotted on rose diagrams, initially keeping the different data sets and lithologies separate. Azimuths are plotted showing the current-to sense, whereas trends are plotted bimodally (*Fig. 2.8*). In most cases, rose diagrams give adequate representation of the dominant, secondary or polymodal paleocurrent directions. If the nature of the amount and data collection allow, then more accurate vector means can be calculated (see *Tucker, 1996*).

2.8 **Examples of plots of different types of orientation directions: (a) unimodal, azimuths of 25 flute cast measurements; (b) bimodal, trend only of 45 groove casts; (c) polymodal, azimuths of current direction from 55 ripple measurements Use overlays on circular graph paper (page 320) for plotting data in the field.**

Stratigraphic procedures and way-up criteria

CORRECT stratigraphic procedure divides rock successions on the basis of lithology (lithostratigraphy), biota (biostratigraphy), time (chronostratigraphy) and key surfaces (sequence stratigraphy). For sedimentary fieldwork, the primary technique is one of descriptive lithostratigraphy by which the succession is divided hierarchically as follows:

Group: several associated/adjacent formations (typically >100 m thick).

Formation: fundamental mappable unit possessing distinctive lithological character and clear boundaries as defined in a designated type section (or sections). Typically 10–100m thick, but thins to 0m.

Member: subdivision of formation having a more uniform lithological character.

Each of these lithostratigraphic units is given a geographical name (with capital letter). Prior to formalizing the stratigraphy, the units may be referred to as lithological units or mapping units. Subsequent work should then be carried out in order to:

- Assign biostratigraphic ages, through the systematic collection and identification of fossil assemblages.
- Determine chronostratigraphic age, through radioisotopic dating of volcanic ash horizons/igneous bodies and so on, stable isotope ratios, or other means.
- Identify key surfaces (unconformities or condensed sequences) along which there is evidence of a significant hiatus, reduction in sedimentation, and/or subaerial exposure, as a preliminary to erecting a regional sequence stratigraphic model.

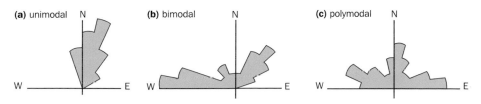

(a) unimodal N W — E **(b)** bimodal N W — E **(c)** polymodal N W — E

cross-stratification (all scales: cross bedding, cross lamination, etc)

truncated tops, curved bases

erosional structures (all scales: channel scours, micro scours, etc)

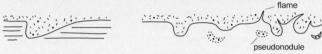

sole marks (grooves, tools, flutes, loads and associated features)

flame

pseudonodule

grading (graded beds/laminae) and structural sequences (Bouma, Stow, etc)

normal grading common but
reverse / symmetrical grading also occurs

surface marks (ripples, desiccation cracks, rain drops, lineation, footprints, etc)

water-escape structures (dishes, burst-throughs, sand volcanoes)

geopetals

spar cement

sediment fill

grazing trails (bed surface)

fossils in growth position eg.stromatolite

cleavage / bedding relationships

2.9 **Criteria for determining stratigraphic way-up of sedimentary strata. Way-up to top of page in each example. Features occur at all different scales; for examples, see relevant field photographs.**

All stratigraphic work depends on the correct recognition of the way-up of strata. This should be routinely checked by means of the way-up criteria summarized in *Fig. 2.9*.

Other techniques

A VARIETY of other field techniques can also be employed to gain maximum sedimentary information from outcrops.

Cleaning and preparing exposures
It is generally important to examine both weathered and fresh surfaces of a rock, using a hammer as necessary to obtain suitable hand specimens. Further techniques that may improve your ability to observe include:

- *Wetting/drying:* some features are best observed when the rock surface is wet, others when dry.
- *Brushing:* a stiff wire or coarse hair brush can be very effective for cleaning lichen, vegetation, soil/mud-wash and so on.
- *Scraping:* poorly consolidated and semi-consolidated sections, in particular, can be much improved by scraping off the surface layer to create a smooth face – a trowel or spade is useful here.
- *Trenching:* softer sediments or those with vegetation cover can sometimes be revealed by trenching or step-trenching up the side of a hill, for example (this may require permission, of course).
- *Other:* I have even seen acid-spray etching of lichen-coated carbonates very effectively used in remote areas. (However, this is not to be recommended on either safety or ecological grounds!).

Sampling and collecting
Where subsequent laboratory work is required (e.g. detailed work on composition, grain size, microfossils, etc.) field samples must be carefully collected, labeled and transported. (Plastic freezer bags are very useful.)

- Plan a sampling strategy in line with the information required – it is often better to return to outcrops at a later time/date solely for collecting samples.
- Collect unweathered *in situ* samples that are representative of the lithology/structure/grain size and so on – their size will depend on the planned analytical work.
- Label the rock sample and the bag with a water-resistant marker pen, adding a way-up arrow as required; for fabric studies both way-up and strike and dip of bedding will be necessary.
- Wrap delicate specimens in newspaper.

Photography
Good clear field photographs are always an invaluable addition to other forms of data recording. This is especially true following the advent of high quality digital cameras and the proliferation of one-hour processing labs for conventional films. Both can be used as real-time field aids when processed or downloaded to the field laptop. They are helpful for: tidying-up field sketches; zooming-in on specific detail, such as subtle trace fossils; preparing broader montages of field relationships; measuring bed thickness variation through vertical sections; and so on.

Remember to record scale, way-up, and photo number in the notebook. For convenience, use the back cover of this field guide (scale and way-up) and write the photo number with a non-permanent marker pen in the space provided. Publication quality photographs will require a little more care and attention in the field: for best lighting conditions (usually a low sun); for the right quality and ASA of colour slide film; or for maximum image resolution with a digital camera.

Laboratory techniques, data manipulation and presentation
Refer to standard texts on sedimentary techniques – e.g. *Bouma (1969), Carver (1971), Lewis (1984), Lindholm (1987), Tucker (1988)*.

PRINCIPAL CHARACTERISTICS OF SEDIMENTARY ROCKS

Pause for thought. Coarse-grained turbidites, central west Chile.

Introduction and facies concept

THE MAIN FEATURES of sedimentary rocks that can be observed directly in the field are:

- Bedding – including thickness, geometry, nature of bed boundaries, dip and strike of the beds, and way-up criteria.
- Sedimentary structures – including erosional, depositional, deformational, biogenic and chemogenic.
- Textures – including grain size, sorting, grading, porosity–permeability, grain morphology, grain surface texture, and sediment fabric.
- Composition – including clast types, mineralogy and fossil content.
- Colour – distinguishing between weathered and fresh surfaces.

These aspects are described briefly in this chapter, by way of summaries, tabulated data and illustrations, and an extensive set of field photographs. Further cross reference is made as appropriate to one or more of the photographs in Chapters 4–14. Each feature can yield different information on the processes and conditions of deposition, such as energy and type of process, current type and flow direction, and depositional environment.

Collectively, these attributes are used to define a sediment facies or lithofacies (lithotype) or, simply, facies. A sediment facies is defined as a sediment (or sedimentary rock) that displays distinctive physical, chemical and/or biological characteristics (of the sort listed above) that make it readily distinguished from the associated facies. These are the primary building blocks of all sedimentological studies. They are used as a means of recording data, for describing sediment outcrops and successions and, subsequently, for making interpretations concerning, for example, depositional processes, environmental conditions or economic significance. These aspects are considered briefly in Chapter 15.

Bedding and lamination

(Figs. 3.1, 3.2; Plates 3.1–3.30, 15.1– 5.32 plus many examples through Chapters 4–14)

BEDDING (thicker than 1cm) and lamination (thinner than 1cm) are the terms used to describe stratification or layering in sediments. Stratification is the generic term that can be used when no specific reference to layer thickness is intended. Lamination is a common internal structure of beds (see page 41). Both laminae and beds are defined by changes in grain size, composition and/or colour that may be more or less distinct. These represent changes in the style of sedimentation caused by different processes, sediment source or depositional environment.

Important features to note include:

- **Bed thickness** and its variation both vertically and laterally. Downcurrent and down-wind decrease in bed thickness is common. Systematic upward increase or decrease in thickness reflect gradual changes in depositional controls. Cyclic variation in thickness reflects a more regular rhythmic pattern of change.
- **Bed geometry** and its scale and continuity. Parallel, continuous bedding indicates stable depositional conditions over the observed area; non-parallel, discontinuous bedding shows local changes have occurred. Wavy and irregular bedding are characteristic of pressure dissolution effects in limestones and dolomites, as well as some rapid, unstable processes of

clastic deposition (e.g. tempestites and some turbidites). Curved and lenticular bedding are common in many environments that show marked lateral variation in depositional process and/or erosion.
- **Bed boundaries** (or contacts), their nature and distinctiveness. Indistinct or diffuse boundaries signify gradual change in depositional regime; sharp, distinct boundaries, the converse. Note that the distinctiveness/sharpness of most bed contacts tends to be enhanced during compaction and diagenesis.
- **Lamination** thickness, geometry and boundaries show exactly analogous features to those of beds.

Observe and measure
1. Bed/lamination thickness and geometry.
2. Bed/lamination boundaries and any bedding plane structures.
3. Any systematic variation in bed/lamination thickness through a vertical succession.
4. Dip and strike of beds – especially where detailed mapping is being undertaken.

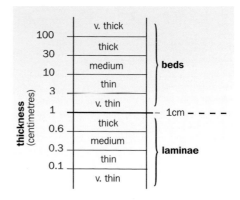

3.1 Thickness and standard terminology for beds and laminae. Some authors use laminae as a subdivision of beds, irrespective of scale, but the most common practice is to use thickness alone as shown here. Stratification, strata, and layer are the terms to use when no specific thickness is implied.

parallel **non-parallel**

planar

planar, parallel

discontinuous,
planar, parallel

planar, non-parallel

discontinuous,
planar, non-parallel

wavy

wavy, parallel

discontinuous,
wavy, parallel

wavy, non-parallel

discontinuous,
wavy, non-parallel

curved

curved, parallel

discontinuous,
curved, parallel

curved, non-parallel

discontinuous,
curved, non-parallel

lenticular

lenticular, sub-
parallel

discontinuous
lenticular, sub-
parallel

lenticular, (channel-like)

irregular

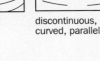

draped, parallel

chaotic (slumped)

irregular

nodular

3.2 **Nature and terminology for geometry of
bedding and lamination. These types can occur at
all different stratification thicknesses.**

Bedding and lamination

3.1   Very thin to very thick-bedded sandstone turbidites; some regular, parallel-sided and with sharp contacts; others lenticular, with several thin sandstone beds passing laterally into a single thick, amalgamated bed (A); note also thickening-upwards (TU) of beds over approximately 8m section.
Eocene, Annot, SE France.

3.2   Thin to medium bedded sandstone–mudstone turbidites. Such regular and parallel-sided bedding is typical of more distal (basin plain) turbidite successions.
Photo by Paul Potter.
Paleocene, Zumaya, N Spain.

3.3   Thin to medium-bedded limestone–marl couplets; more or less regular, sub-parallel, with marked thinning as beds drape over small *in situ* mounded bioherm (B); limestone–marl contacts mainly sharp but wavy/irregular due to inter-stratal dissolution. Height of roadcut section 8m.
Miocene, S Cyprus.

3.4 Lenticular or wedge-shaped conglomerate bed within mainly sandstone-rich fan-delta succession; sharp bedding contacts, erosive at base, suggesting small-scale channel-fill deposit.
Miocene, Pohang Basin, SE Korea.

3.3 Part of megabed – approximately 60m thick, slide–debrite–turbidite composite unit – sub-vertical bedding, as picked out by bed-parallel alignment of elongate clasts; top to right.
Miocene, Tabernas Basin (Gordo Megabed), SE Spain.

3.6 Thick and very thick-bedded turbidite sandstones/pebbly sandstones; beds structureless or with slight normal grading, and so indistinct except where picked out by clast-rich base. Geologist (for scale) is sitting on prominent joint surface; bedding dips steeply from upper left to lower right.
Triassic, central west Chile.

3.7 Thin to medium bedded volcaniclastic succession; sub-parallel bedding with contacts mainly sharp, some gradational. Dark beds are coarse-grained, scoriaceous fall deposits; pale beds are fine-grained pumiceous/biogenic hemi-pelagites (see Chapter 14). Note minor faults to left of geologist.
Miocene, Miura Basin, south central Japan.

3.8 Siltstone/fine sandstone laminae, lenticular and wavy laminae and very thin beds within mudstones; lenticular (L) to flaser (F) lamination in parts, and some siltstone laminae dis-continuous; most laminae have sharp contacts; shallow-marine tidal setting. Coin 2.5cm.
Triassic–Jurassic, Los Molles, west central Chile.

Erosional structures

THESE STRUCTURES are formed by the erosive action of currents, effected by their sediment load. They commonly occur as casts on the underside of beds (sole marks) or depression marks on the tops of beds (surface marks) and as inclined or (sub-)planar scours through a sedimentary unit. They include:

Small and medium-scale erosional structures
(Fig. 3.3; Plates 3.9–3.22, 5.11, 5.16, 15.6)

- *Rill marks* form as a small-scale dendritic channel network due to water runoff over sand or silt-mud slopes that are periodically exposed subaerially (e.g. beaches, tidal slopes and so on). They are rarely preserved in ancient rocks.
- *Furrows and ridges* (or longitudinal scours) are formed under low velocity currents mainly by grain reorganization on the bed surface, and minor erosion along flow-parallel furrows.
- *Tool marks* are formed when objects being carried by currents impact or scrape along the sediment surface. Discontinuous tool marks include prod, skip and bounce marks of various shapes and sizes. More continuous tool marks are elongate grooves and chevrons, both oriented parallel to flow and with the V-shaped crenulations closing downstream.
- *Obstacle scours* are crescent or horseshoe-shaped depressions that form around a larger stationary obstacle (e.g. pebble, wood/fossil fragment) on the bed surface.
- *Flutes* are very similar to obstacle scours but form as a result of fluid erosion without an obstacle, although some bed surface irregularity may initiate flute development. They occur in a variety of shapes, from narrow elongate to broad transverse scours, a wide range of scales, and either as isolated marks or in distinct clusters.

- *Gutter casts* are isolated, elongate, U- or V-shaped depressions that are slightly to moderately sinuous over several metres downstream. They result from a combination of fluid scour and the erosive action of coarser grains within the flow.
- *Other scours* form as less regular erosion surfaces cutting down from the base of beds, or within beds. They may result from short-term changes in flow conditions, from sliding and slumping during or after a flow event (slide scars), or from minor channel erosion.
- *Wind-erosion mounds*, showing a positive relief on the bed surface, are the erosional remnants left as strong winds flow over a damp, slightly cohesive surface. They are similar in form (but opposite in relief) to flutes, with a blunt upwind nose and tail streaked out downwind.
- *Glacial striations and pluck marks*, are formed by the scouring and erosive action of debris carried in flowing ice.

Observe and measure:
1. Shape in three-dimensions as far as possible from the exposure – e.g. complex or simple form, symmetrical or asymmetrical, regular form (as above) or irregular.
2. Dimensions in true or apparent section – e.g. depth, width, length.
3. Orientation and sense of flow direction where this can be determined.
4. Cross-cutting relationship where two or more erosional and/or other structures occur – determine chronology of events.
5. Way-up of strata, where this can be determined (see page 26).
6. Evidence of planar erosion where no scour feature is present – e.g. a bed-parallel string of shale-clasts within a sandstone unit; intensively bored and burrowed bedding surfaces; burrows filled with a sediment type not otherwise present in the succession; hard grounds cemented with $CaCO_3$ or with Fe–Mn encrustation; hiatuses in the faunal or floral succession.

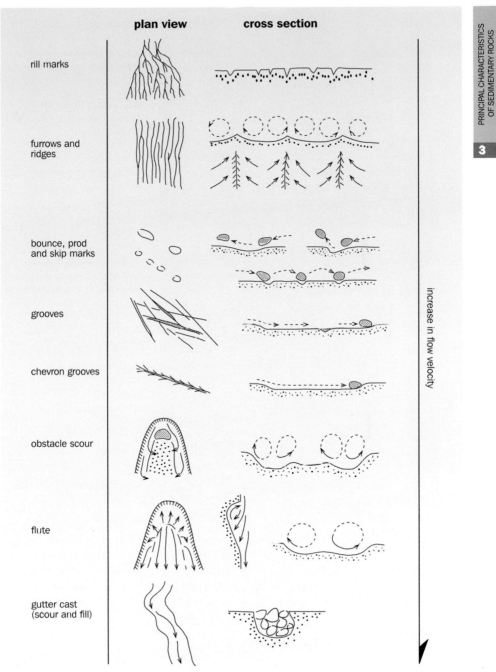

3.3 Small and medium-scale erosional features and their approximate relationship to flow velocity.

Erosional structures

3.9 Rill marks over surface of limestone caused by the chemical erosive action of rain-water (weathering).
Wine pouch 15cm wide.
Cretaceous limestones, Recent rill marks, SE Spain.

3.10 Glacial striae and pluck marks on smoothed glacial pavement, caused by the erosive action of debris contained in the basal layers of a glacier. Darker cracks are joints. Width of view 1m.
Carboniferous glacial episode scouring Precambrian rock surface, Hallett Cove, S Australia.

3.11 Detail of 3.10.

3.12 Large-scale glacial striae and associated glacial scour features on an Ordovician sandstone surface. The ice sheet probably moved across an exposed or very shallow shelf. Width of view in foreground 12m.
Photo by Paul Potter.
Ordovician, Tassili, S Algeria.

3.13 Raindrop impressions preserved in intertidal mudstone. A rare find in the sedimentological record. Bedding plane, width of view 12cm.
Permo-Triassic, Scotland, UK.

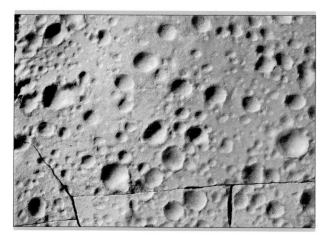

3.14 Surface view of faint lineation (extremely low-amplitude ridges and furrows) on bedding plane, caused by lower plane bed flow-regime running water. Width of view 30cm.
Early Cambrian, Flinders Ranges, S Australia.

3.15 Surface view of sandstone bed with early formed asymmetrical ripples, now partially obscured by current lineation (ridges and furrows) oriented approximately along ripple crest strike (R); shallow-marine, probable tidal influence.
Lens cap 6cm.
Silurian, Quebrada Ancha, NW Argentina.

3.16 Surface view of turbidite bed with parallel erosional grooves and irregular-shaped scour marks (possible prod and bounce marks). Cross-cutting joints from bottom left to upper right. Lens cap 6cm.
Carboniferous, near San Juan, NW Argentina.

3.17 View of basal surface of turbidite bed, showing erosional scours (some flute-shaped, as marked) and loads; current direction probably from left to right. Width of view 40cm.
Miocene, SE Cephallonia, Greece.

3.18 Flute marks as casts clustered on base of turbidite sandstone bed. Current direction from right to left, determined from flute orientation and geometry (narrow and deep upflow, broader and shallower downflow). Bedding plane (base), width of view 20cm. *Ordovician, Southern Uplands, Scotland.*

3.19 Cross-section of large erosional flute cut by a calcirudite turbidite into calcilutites; deepwater succession. View approximately at right angle to flow direction. Lens cap 6cm. *Cretaceous, Monte Pietralata, central Italy.*

3.20 Thick succession of lacustrine delta-front/delta-slope siliciclastic sediments showing several large-scale, low-angle erosional surfaces (dashed lines); these features were cut by slides or by channel erosion and rapidly filled by prograding units, leading to the large-scale horizontal to cross-stratified deposits observed. Vertical lines and colour streaks result from secondary weathering.
Width of view 15m.
Central Hokkaido, N Japan.

3.21 Sharp, steep erosional margin (dashed line) of small-scale deep-water channel, cut into thin-bedded turbidites and filled by conglomeratic debrite over sandy turbidite or slurry bed. Width of view 4m.
Miocene, Tabernas Basin, SE Spain.

3.22 Broad erosional scours at base of sandstone and pebbly sandstone turbidites. Note several sets of erosional scours (dashed lines).
Width of view 5m.
Cretaceous, S California, USA.

Large-scale erosional features
(Plates 3.21, 3.22, 15.9–15.11, 15.14–15.15)

Most of the outcrop-scale erosional features noted above also occur at a much larger scale. Detailed seafloor mapping has shown furrows and ridges that are 10m wide and 10km long, and megaflutes 500m wide and 5km long.

- *Planar erosion surfaces* can be very widespread – particularly hiatuses, which may occur over wide tracts of seafloor.
- *Slide/slump scars* occur at all scales up to 1000km² in area. They are smooth, concave-upward surfaces with a steep headwall.
- *Channels* also occur at a wide variety of scales. It is important to remember that what appears channel-like in outcrop may, in fact, have been formed by some other erosional process (e.g. megaflute scour, slump event), or may be the thalweg of a much larger channel.
- *Unconformities* (see Chapter 15, page 263)

Depositional structures

THESE STRUCTURES are formed during the deposition of sedimentary material by one of a large number of processes, both subaerial and subaqueous. They occur most commonly on bedding surfaces and within beds.

Parallel lamination/stratification
(Fig. 3.4; Plates 3.23–3.30)

This occurs in most lithologies and most environments, and is defined by variation in grain size, composition and/or colour. Variations relate to flow velocity, grain size and type of sediment. Lamination is the term used for fine layering (<1cm), and stratification for coarser layering, or when no particular thickness is implied.

- Crude parallel (and subparallel) stratification in gravels and pyroclastic surge deposits (see also *Plates 4.4–4.11, 5.5*).
- Crude parallel stratification in muddy sandstones can form by slurry-flows (intermediate between debris flows and turbidity currents).
- Upper phase plane lamination in sands and silts, with parting lineation typical on bed surfaces (see also *Plate 5.5*).
- Lower phase plane lamination in coarse sands (see also *Plate 5.3, 5.4, 5.11, 5.16*).
- Silt–mud (or fine sand–mud) lamination, either planar–continuous or streaky–discontinuous (see also Chapter 6).
- Varve lamination (graded silt–mud couplets ± organic matter) typical of lake sediments.
- Evaporite lamination, comprising different mineral phases ± traces of organic matter (see also *Plates 11.5–11.7*), and including chemical lamination formed in hot spring precipitates, tufas and karstic speleothems (*Plates 7.27–7.30*).
- Fissile lamination (very fine-scale – parallel, wavy, anastomosing) typical of some clays; organic-carbon-rich sediments (black shales), and micaceous sediments (see *Plates 6.20, 7.8*).
- Microbial (algal) lamination in limestones – planar, undulatory and crenulated (see also *Plates 7.7, 7.16–7.19*).

Observe and measure:
1. Continuity and lateral relationships.
2. Cause or type of lamination – e.g. grain alignment, grain size, composition, colour, primary growth.
3. Thickness of stratification.
4. Internal grading of individual laminae (normal or reverse).
5. Any grouping, cyclicity or rhythmicity of lamination, that may suggest a longer-term control on deposition.
6. Relationship of laminated unit to any sequence of structures – e.g. Bouma or Stow sequence (page 60/graded beds).

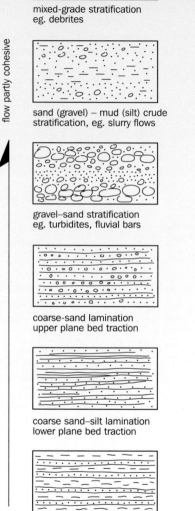

flow partly cohesive

increase in flow velocity

mixed-grade stratification
eg. debrites

sand (gravel) – mud (silt) crude
stratification, eg. slurry flows

gravel–sand stratification
eg. turbidites, fluvial bars

coarse-sand lamination
upper plane bed traction

coarse sand–silt lamination
lower plane bed traction

silt–mud lamination
suspension deposit with
depositional shear sorting

low velocity

silt–mud streaky lamination
suspension + depositional
shear sorting

varve lamination
suspension fall-out

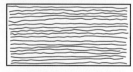

fissile lamination
suspension fall-out + organic
matter

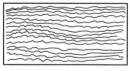

evaporite lamination
evaporitic precipitation
+ bio/chemical variation

microbial (algal) lamination,
sub-parallel

microbial (algal) lamination,
crenulated

suspension fall-out

**evaporitic
precipitation**

biogenic precipitation

3.4 Different types and origins of parallel stratification and lamination.

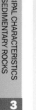

Parallel lamination and stratification

3.23 Crude parallel to sub-parallel stratification in conglomerates and pebbly sandstones, deep-water turbidite succession. Note oscillation grading and no clear bed boundaries. Width of view 2m.
Cretaceous, S California, USA.

3.24 Crude parallel to sub-parallel stratification in sandstones and pebbly sandstones, fluvial–shallow marine succession. Note oscillation grading and no clear bed boundaries. Width of view 1.2m.
Permian, Bondi Beach, Sydney, Australia.

3.25 Crude parallel to sub-parallel stratification in pyroclastic surge deposit. Note also the autobrecciated imbricated clast horizon across centre of view. Coin 2.5cm.
Quaternary, Tenerife, Spain.

3.26 Planar parallel lamination in volcaniclastic turbidite, formed during upper-plane bed flow phase (Bouma B division) in coarse sand-sized material; part of very thick bed with slight normal grading. Note also large floating clast (F) to lower left. Lens cap 6cm.
Carboniferous, Rio Tinto, S Spain.

3.27 Planar parallel lamination in fine–medium sandstone, formed by lower-plane bed flow phase in probable beach foreshore setting. Note also low-angle erosional scours (reactivation surfaces, R), probably caused by storm event; also minor vertical fault left of centre. Hammer 25cm.
Silurian, near Canberra, Australia.

3.28 Interbedded siltstone–
mudstone laminated interval,
with partial Stow turbidite
sequences (arrows), disrupted
slide-slump unit (S), and thin-
bedded parallel-laminated
siltstone/sandstone turbidites
(lower-plane bed phase); minor
cross-lamination in parts;
lacustrine basin.
Hammer 30cm.
*Triassic, Puquen, west central
Chile.*

3.29 Lenticular and wavy
lamination in calcarenite–
calcilutite succession, in part
caused by inter-stratal dissolution and in part by deposition
as contourites. Lenses represent
current-winnowed coarse
fraction
Width of view 10cm.
Paleogene, southern Cyprus.

3.30 Microbial/evaporite
lamination as prominent thin
white layers within a lacustrine
silty mudstone. Hammer 25cm.
*Plio–Pleistocene, Goyder, Lake Eyre,
Australia.*

disrupted lamination,
very rapid deposition
and collapse of
micro-ripple structure

wispy lamination
discontinuous streaks
and lenses

Wavy, lenticular and flaser lamination
(Fig. 3.5; Plates 3.36, 3.62–3.65, 6.12, 6.21,
11.5–11.7)

This occurs mainly as intermediate structures
(i.e. part parallel- and part cross-lamination)
in mixed grade sediments (e.g. interbedded
sand–mud or limestone–marl facies). Variations relate principally to flow velocity and
grain size.

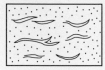

flaser lamination
excess sand/silt
deposition with minor
mud

lenticular lamination,
continuous–
discontinuous
no internal structures,
rare micro-cross
lamination

- *Wispy lamination* is a very fine-scale,
  discontinuous, wavy form, mainly silt
  wisps in a background of mud.
- *Lenticular lamination* comprises individual silt/sand lenses, typically with internal
  (micro) cross-lamination, within a fine-
  grained (mud) unit.
- *Wavy lamination* is made up of intercon-
  nected and overlapping lenses of silt/sand
  within a mud unit. Where lenses display
  internal structure this is typically either
  of fading ripples, or of long wavelength
  low-amplitude ripples (ie very low-relief
  ripple forms, preserving the silty stoss
  side as it climbs slightly over a subtle
  muddy trough).
- *Flaser lamination* comprises wavy or
  lenticular silts and fine sands with thin
  wavy to wispy partings of mud.
- *Contorted lamination* can form as a disrupted primary structure due to rapid
  sedimentation, as well as by post-depositional deformation (see also page 76,
  Fig.3.15).

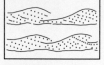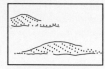

fading ripples, micro-
cross lamination with
muddy troughs

isolated fading ripples
(starved ripples)

low-amplitude long-
wavelength ripples
barely perceptible
micro-cross lamination
+ muddy troughs

wavy lamination with
no internal structure

Observe and measure
As for parallel and cross-lamination.

3.5 Wavy, lenticular, and flaser lamination
typical of interlayered (heterolithic) fine-grained
sediments (silt/sand and mud grades).

Wavy/lenticular lamination
Cross-lamination/stratification

3.31 Plan view of sharp-crested, sinuous, eolian ripples (showing slight asymmetry), passing into smooth wind-scoured surface.
Width of view 1.2m.
Present day, Sahara desert, S Tunisia.

3.32 Plan view of sinuous, long-wavelength, asymmetrical ripples formed by high velocity tidal current. Note coarse shell and pebble lag in ripple troughs, and weak surface lineation. Flow from right to left.
Hammer 45 cm.
Present day, La Rance estuary, N France.

3.33 Bed surface view of round-crested, symmetrical ripples, formed by wave action. Note ladder-like interference ripples (tidal) to right and more irregular interference pattern to left. Scattered shell debris.
Hammer 45cm.
Present-day, La Rance estuary, N France.

3.34 Bedding plane view of round-crested symmetrical wave ripples preserved on surface of oolitic limestone bed. Shallow-marine environment.
Hammer 25cm.
Ordovician, West Yunnan, SW China.

3.35 Bedding-plane view of sandstone bed with round-crested symmetrical wave ripples. Fallen block to right shows evidence of interference ripples typical of shallow marine (tidal) environment.
Hammer 25cm.
Jurassic, near Whitby, NE England.

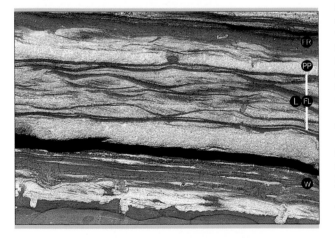

3.36 Fine-grained turbidites showing planar parallel (PP), wavy (W), lenticular (L), and flaser lamination (FL). Note micro cross-lamination in many of the silt lenses, some of these showing fading ripple structure (FR), in which the silt laminae (light-coloured) on the ripple crests fade into mud laminae (dark) in the troughs. Also note sharp bases, with micro-scouring and some load and flame structures (bottom).
Width of view 15cm.
Paleogene, California, USA.

3.37 Stacked symmetrical wave-ripple cross-lamination with some planar parallel lamination; oscillatory flow dominant in shallow-marine fine-grained sandstone succession, with episodic erosive (possible storm) events and planar laminae deposition. Width of view 16cm.
Pliocene, Sorbas Basin, SE Spain.

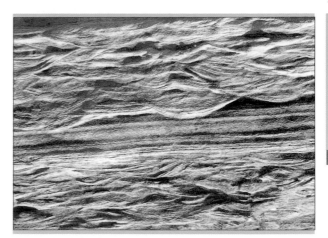

3.38 Herringbone (bi-directional) cross-lamination in fine-grained sandstones (centre of view); typical of alternating flow directions (arrows) in tidal environment. Coin 2.5cm.
Pliocene, Sorbas Basin, SE Spain.

3.39 Climbing-ripple cross-laminated sandstone turbidite unit (CR), overlying planar parallel laminated turbidite succession; darker layers are finer grained (muddy/micaceous) material; flow to right. This is typical of very rapid deposition from sediment-laden currents. Width of view 50cm.
Eocene, near Annot, SE France.

3.40 Cross-laminated sandstone showing tabular cross-bedding with curved bases and sharp erosive tops, flow to left. Width of view 80cm. *Pliocene, Sorbas Basin, SE Spain.*

3.41 Interbedded sandstones and siltstone–mudstone units. The brown sandstone bed above the hammer shows large-scale swaley cross-stratification. (SCS), comprising mainly broad, concave-up laminae. Formed by high-energy storm waves; related to HCS (below). Hammer 30cm. *Jurassic, Osmington Mills, S England.*

3.42 Hummocky cross-stratification (HCS) in shallow-marine sandstones. Gently undulating low-angle hummocks and concave-up swales. Formed by high-energy oscillatory to complex flow patterns typical of storm-waves. Hammer 25cm. *Silurian, near Canberra, Australia.*

3.43 Large-scale cross-stratified sets (sets up to 3.5m) in bioclastic calcarenite; probable shallow-marine, tidal environment. Coarser-grained calcarenite (light-coloured) layers represent high-energy flow; thin, finer-grained calcilutite (darker-coloured) layers represent low-energy deposits. Typical of peak tidal flow to slack tidal period 'mud drape' deposition.
Note that the wavy surfaces of individual foreset bedding is due to inter-stratal dissolution.
Width of view 4m.
Cretaceous, Bonafaccio, Corsica, France.

3.44 Large-scale eolian dune cross-stratification in weakly cemented 40m high, present-day dune field. Width of view 1.2m.
Recent, Sahara Desert, S Tunisia.

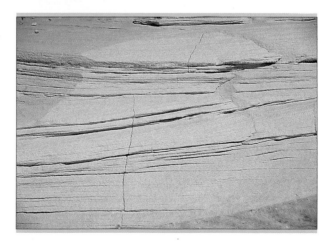

3.45 Large-scale, wedge-shaped, eolian dune cross-stratification, typical of wind-blown processes. Width of view 2.5m. Photo by Paul Potter.
Cretaceous, Rio Claro, Sao Paulo State, Brazil.

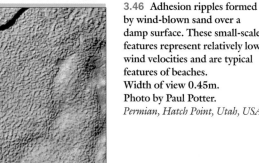

3.46 Adhesion ripples formed by wind-blown sand over a damp surface. These small-scale features represent relatively low wind velocities and are typical features of beaches.
Width of view 0.45m.
Photo by Paul Potter.
Permian, Hatch Point, Utah, USA.

3.47 Large-scale cross-stratification in point bar sands, revealed by digging long trenches into large point bars at low water along the Mississippi River near New Madrid Bend. Courtesy of Sedimentology–Stratigraphy field class from the University of Cincinnati.
Photo by Paul Potter.
Recent, Fulton Co., Kentucky, USA.

3.48 Partly eroded sand bar in modern river showing dominant planar sub-parallel stratification, local gravel concentrations, and zones of small-scale ripples. Width of view 3.5m.
Recent, northern Sicily, Italy.

3.49 Low-relief gravel bar on modern dry river bed, with parallel cross-bar gravel ridges (equivalent to sand ripples). Width of view 3.5m.
Recent, Talacasto, NW Argentina.

3.50 Large-scale lozenge-shaped gravel bar with partly vegetated surface in modern river. Bar length 30m.
Recent, southern Cyprus.

Cross-lamination/bedding/stratification

(*Plates* 3.31–3.50, 5.6–5.14, 7.26, 11.7, 14.29, 15.41–15.44)

This is widespread in most lithologies and environments. It results from the deposition, migration, and preservation of inclined lamination that develops as the internal structure within many types of surface bedform. These include: micro-ripples (lenses – as above), ripples, sandwaves, dunes, antidunes, hummocks, and bars. Where bedding planes are exposed, the smaller-scale bedforms can be recognized on the bed surface, and measurements made (*Fig. 3.6, 3.7*). Larger-scale bedforms are generally recognized from their internal cross-stratification alone. The principal types are classified according to the scale of the feature, which is related to grain size, flow velocity and depth. Variations then occur, dependent on the type of current and the depositional setting (*Fig. 3.11*). A cross-stratified bed may comprise a single set of cross-laminae or -bedding, or numerous sets, known as a coset. In the same way, as for parallel stratification, use the term 'cross-lamination' where the individual cross-strata are

3.6 Terminology and scales for different current-induced bedforms (ripples, dunes, sandwaves, bars, and draas).

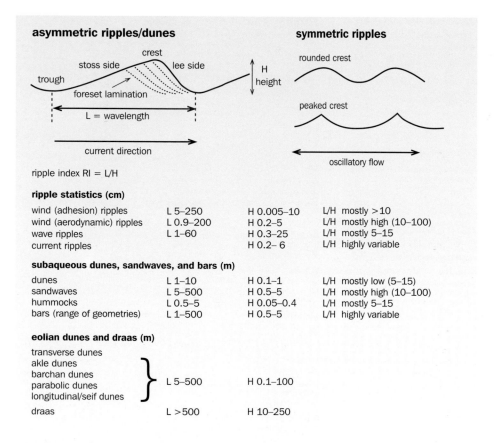

asymmetric ripples/dunes

crest
stoss side — lee side
trough
foreset lamination
L = wavelength
H height

current direction

ripple index RI = L/H

symmetric ripples

rounded crest

peaked crest

oscillatory flow

ripple statistics (cm)

	L	H	L/H
wind (adhesion) ripples	L 5–250	H 0.005–10	L/H mostly >10
wind (aerodynamic) ripples	L 0.9–200	H 0.2–5	L/H mostly high (10–100)
wave ripples	L 1–60	H 0.3–25	L/H mostly 5–15
current ripples		H 0.2–6	L/H highly variable

subaqueous dunes, sandwaves, and bars (m)

	L	H	L/H
dunes	L 1–10	H 0.1–1	L/H mostly low (5–15)
sandwaves	L 5–500	H 0.5–5	L/H mostly high (10–100)
hummocks	L 0.5–5	H 0.05–0.4	L/H mostly 5–15
bars (range of geometries)	L 1–500	H 0.5–5	L/H highly variable

eolian dunes and draas (m)

	L	H
transverse dunes akle dunes barchan dunes parabolic dunes longitudinal/seif dunes	L 5–500	H 0.1–100
draas	L >500	H 10–250

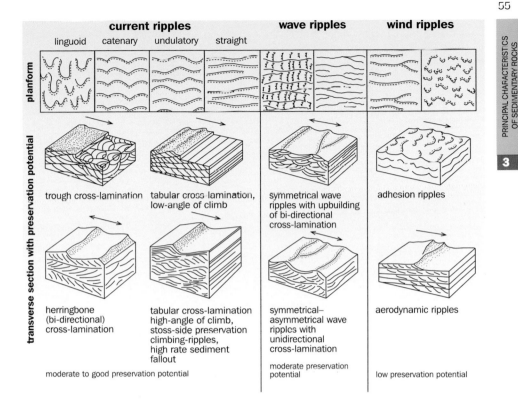

3.7 Small-scale cross-lamination formed by wind, wave and current ripples. Lee sides (steeper, downstream-facing) are stippled. Bi-directional forms are stippled on both sides (except for wave ripples in planform).

<1cm thick, 'cross-bedding' where the individual cross-strata are >1cm, and 'cross-stratification' for no implied layer thickness.

- *Very small-scale cross-lamination* – lenticular, wavy and flaser lamination (page 46, *Fig. 3.5*).
- *Small-scale (ripple) cross-lamination* (*Fig. 3.7*) (set height <10cm, cross-laminae thickness ~ few mm) includes: wind ripples, wave ripples, current ripples; various planforms; tabular or trough cross-sectional forms; typical downflow migration, but also reverse-flow and interference ripples in bi-

directional tidal systems and reverse flow ripples in the troughs of larger bedforms (e.g. dunes); climbing-ripple sets where the deposition rate is high.

- *Medium-scale (dune) cross-bedding* (*Fig. 3.8*) (set height >10cm <1m, cross-bed thickness ~ many mm) includes: small eolian dunes, current-formed sandwaves, dunes and antidunes, storm-generated hummocky cross-stratification (HCS) and swaley cross-stratification (SCS); wide variety of planforms and cross-sectional forms.
- *Large-scale cross-bedding* (*Fig. 3.9, 3.10*) (set height >1m, cross-bed thickness ~ mm to cm) includes: variety of eolian dunes and draas; high-energy river-bed bars, sand waves, epsilon cross-bedding and Gilbert-type cross-bedding; variety of planforms and simple to highly complex cross-sectional forms.

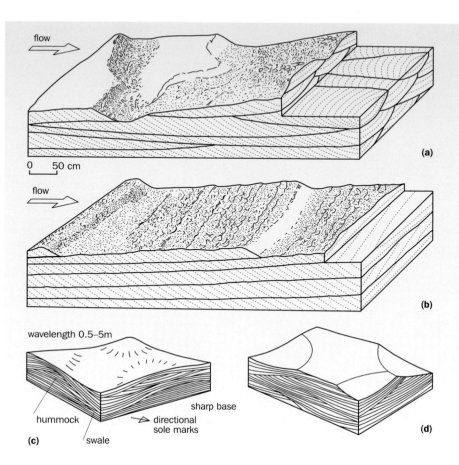

3.8 **Medium-scale cross-stratification formed by the migration of (a) dunes (trough cross-bedding), (b) sandwaves (tabular cross-bedding); and by storm-generated bedforms – (c) hummocky cross-stratification, (d) swaley cross-stratification.**
Modified after Collinson & Thompson (1989) – (a, b); and Tucker (1996) – (c, d).

Observe and measure:

1. Ascertain scale of feature – thickness of sets/cosets, thickness of cross-strata (laminae or beds).
2. Determine shape of cross-stratified sets – tabular, trough, wedge or more complex.
3. Measure direction and maximum angle of dip of foresets – N.B. several readings necessary for paleocurrent analysis.
4. Note maximum/mean grain-size, and grading or sorting.
5. Determine style of cross-stratification (*Figs. 3.6–3.10*) – infer current type, energy and depositional setting (*Fig. 3.11*).

Graded beds and structural sequences

(*Plates* **3.23, 3.24, 3.51–3.65, 5.1, 5.16, 5.18, 6.5, 6.15, 6.20, 8.10, 11.8, 14.16, 14.22, 14.28**)

Some form of grading is very common through individual laminae, parts of beds and whole beds in many different rock types and depositional environments. In most cases this involves systematic grain-size changes – normal (or positive) indicates an upward decrease in size and is the most common, and

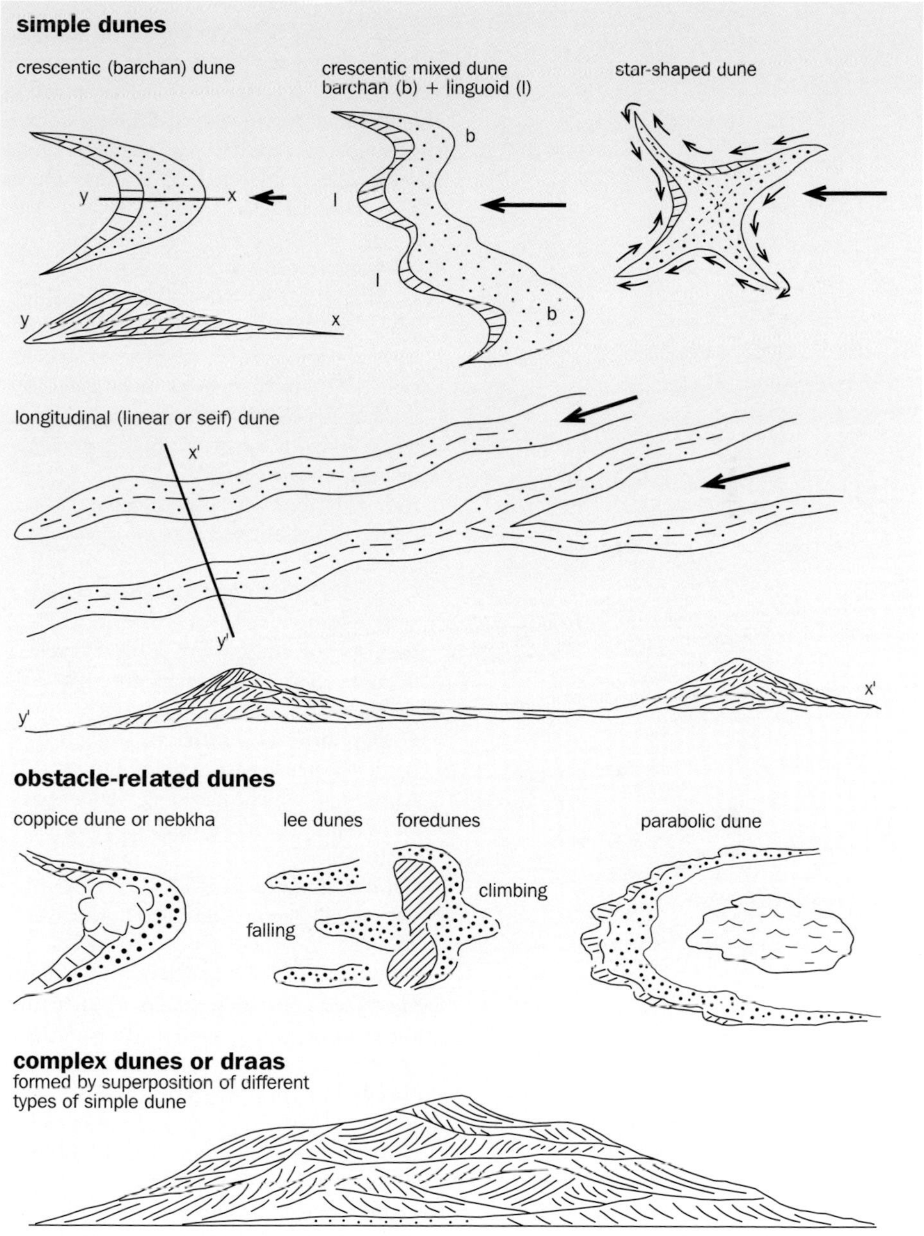

simple dunes

crescentic (barchan) dune

crescentic mixed dune
barchan (b) + linguoid (l)

star-shaped dune

longitudinal (linear or seif) dune

obstacle-related dunes

coppice dune or nebkha

lee dunes foredunes

falling climbing

parabolic dune

complex dunes or draas
formed by superposition of different
types of simple dune

3.9 Large and very large-scale bedforms and associated cross-stratification formed by eolian processes. Although not entirely satisfactory or complete as a classification scheme, eolian dunes are generally referred to as simple, obstacle-related, or complex in form.

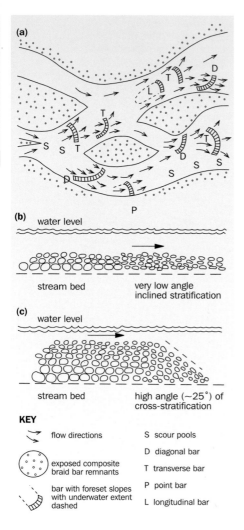

KEY

⤳ flow directions

○ exposed composite braid bar remnants

⤸ bar with foreset slopes with underwater extent dashed

S scour pools

D diagonal bar

T transverse bar

P point bar

L longitudinal bar

3.10 Parallel, inclined, and cross-stratification in fluvial bars in braided gravelly streams: (a) gravel bar types, (b) cross-section of longitudinal/diagonal bar, (c) cross-section of transverse bar. Bars develop as a diffuse gravel core stops moving, forming a lag; this grows upwards and downstream. Note normal vertical and lateral grading. *Modified after Collinson & Thompson (1989).*

reverse (or negative) indicates an upward increase. Grain-size changes are commonly accompanied by changes in composition and/or colour, and in some cases by systematic changes in sedimentary structures (see below). The main types of grading are as follows: (*Fig. 3.12*)

- *Normal graded bed* (also positive grading): can show distribution grading, where the whole grain-size distribution fines upwards, coarse-tail grading, where only the coarsest particles show notable grading, base-only and top-only grading, where only the lower or upper parts respectively are graded.
- *Reverse graded bed* (also negative grading): variations as for normal grading.
- *Composite graded bed:* can show symmetrical grading (i.e. normal-to-reverse), oscillation grading (with repeated cycles), or more complex grading through a single bed.
- *Graded laminated unit:* a distinct bed made up of alternating silt and mud laminae, in which successive silt laminae show a progressive upward decrease in grain size.
- *Graded laminae:* individual laminae can be normally graded or reverse graded as for beds.

Standard sequences of sedimentary structures through individual graded beds commonly result from sediment-charged waning flows – e.g. turbidity currents, storm surges, river floods and pyroclastic flows. As the flow velocity slows, so different bedforms are generated and preserved sequentially as internal structures. The complete or idealized sequence is rarely found but partial sequences are common, showing the correct order of divisions except where velocity fluctuation or reversal has occurred. Contourites show structural sequences that result from a long-term increase then decrease in flow velocity.

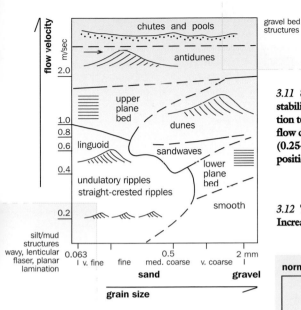

3.11 Sedimentary structure matrix showing stability fields for subaqueous bedforms in relation to specific velocity–grain size conditions. The flow depth for this diagram is relatively shallow (0.25–0.5 m); varying flow depth will change the position of the partitions between fields.

3.12 Types of grading typical in sediments. Increasing grain size to right.

Sequences include:
(*Fig. 3.13*)
- *Stow sequence* – for fine turbidites.
- *Bouma sequence* – for medium turbidites.
- *Lowe sequence* – for coarse turbidites.
- *Dott–Bourgeois sequence* – for storm deposits.
- *Sparks sequence* – for ignimbrites.
- *Stow–Faugères sequence* – for contourites.

Other sequences are currently being developed for debrites and hyperpycnites.

Observe and measure:
1. First determine bed boundaries and hence whether the grading is through the whole or only part of a bed.
2. Characterize the type of grading present.
3. Measure variation in maximum and mean grain size.
4. Look for associated composition/colour grading.
5. Note the presence of any structural sequences, including the divisions that occur.
6. Remember that not every graded bed fits one model. Think before you interpret.

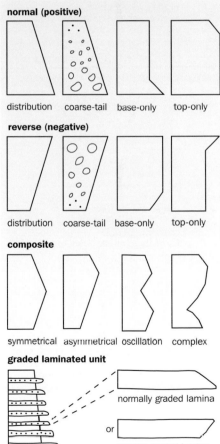

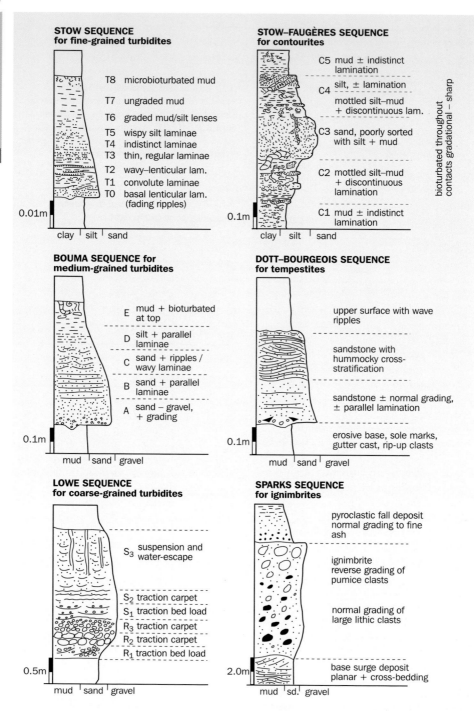

STOW SEQUENCE
for fine-grained turbidites

T8 microbioturbated mud

T7 ungraded mud

T6 graded mud/silt lenses
T5 wispy silt laminae
T4 indistinct laminae
T3 thin, regular laminae
T2 wavy–lenticular lam.
T1 convolute laminae
T0 basal lenticular lam.
(fading ripples)

0.01m

clay | silt | sand

STOW–FAUGÈRES SEQUENCE
for contourites

C5 mud ± indistinct lamination

C4 silt, ± lamination

mottled silt–mud + discontinuous lam.

C3 sand, poorly sorted with silt + mud

C2 mottled silt–mud + discontinuous lamination

C1 mud ± indistinct lamination

0.1m

clay | silt | sand

bioturbated throughout
contacts gradational – sharp

BOUMA SEQUENCE for
medium-grained turbidites

E mud + bioturbated at top

D silt + parallel laminae

C sand + ripples / wavy laminae

B sand + parallel laminae

A sand – gravel, + grading

0.1m

mud | sand | gravel

DOTT–BOURGEOIS SEQUENCE
for tempestites

upper surface with wave ripples

sandstone with hummocky cross-stratification

sandstone ± normal grading, ± parallel lamination

erosive base, sole marks, gutter cast, rip-up clasts

0.1m

mud | sand | gravel

LOWE SEQUENCE
for coarse-grained turbidites

S3 suspension and water-escape

S2 traction carpet
S1 traction bed load
R3 traction carpet
R2 traction carpet
R1 traction bed load

0.5m

mud | sand | gravel

SPARKS SEQUENCE
for ignimbrites

pyroclastic fall deposit normal grading to fine ash

ignimbrite reverse grading of pumice clasts

normal grading of large lithic clasts

base surge deposit planar + cross-bedding

2.0m

mud | sd. | gravel

3.13 **Standard sequences of sedimentary structures that result from different depositional processes.** *Synthesized from original sources.*

Structureless (or massive) beds

(*Plates* **3.6, 4.2– 4.5, 4.11–4.18, 5.2, 5.17, 6.8, 6.11, 6.19, 7.4, 7.21, 7.31–7.35, 8.10, 14.13**)

Many beds and parts of beds have no internal primary structures and no grading. Careful examination is required to ascertain that this really is the case, as surface weathering and uniform grain size can obscure subtle structures. Mudstones and micrites, in particular, can appear structureless simply because they are badly weathered.

Truly structureless beds – (I prefer this to the term massive to avoid confusion with beds having very great thickness) – are the result of (a) no primary structures forming during deposition, or (b) the post-depositional obliteration of primary structures. The former generally arises through very rapid sedimentation or 'dumping' so that there is insufficient time to develop structures or even grading. This is typical of some turbidites, river flood and volcaniclastic sediments, as well as most rock avalanche deposits, debrites, and glacial tillites. Rapidly deposited turbidite or river-flood sands may be structureless apart from the development of water-escape features (page 77; *Plates* **3.89–3.97**).

The obliteration of primary structures and hence the development of a structureless bed can result from very thorough bioturbation or churning, wholesale dewatering, and diagenetic recrystallization. Where there has been wholesale disruption and/or mixing of different beds, then a chaotic structureless fabric is developed (see below, *Chaotica*).

Graded beds and structural sequences

3.51 Part of 2.5m thick, normally-graded bed (arrow), from pebbly sandstone (bottom right) to fine sandstone (top left); no other structures evident; deep-water turbidite succession.
Miocene, Tabernas Basin, SE Spain.

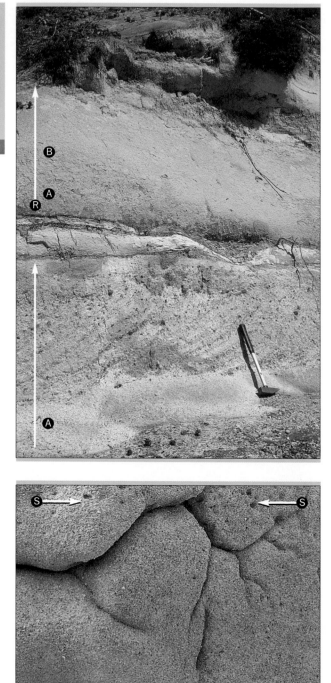

3.52 Two thick sandstone beds within turbidite succession. Upper bed shows reverse-graded base (R) followed by normally graded sandstone (arrow), with Bouma divisions AB as marked. Lower bed shows normal grading from pebbly sandstone to sandstone (Bouma Division A) capped by dune cross-bedded sandstone. This is an unusual development in turbidite systems, which probably indicates a prolonged phase of high energy flow, tractional transport and bed reworking.
Hammer 45 cm.
Cretaceous, S California, USA.

3.53 Reverse-graded base of very thick (15m) mega-turbidite, slightly erosive over mudstone. Small cavities (S), that become more common upwards, remain from shale clasts that have been mainly picked out by coastal action and weathering.
Width of view 60cm.
Oligocene, Reitano, northern Sicily, Italy.

3.54 Very thick-bedded pebbly sandstone to sandstone turbidite showing reverse (R) to normal (arrow) grading. Bouma divisions (AB) as marked. Note that various cracks are due to weathering and jointing. Base of turbidite, marked with dotted line, is also a horizontal joint – minor fault plane. Lens cap 6cm. *Cretaceous, S California, USA.*

3.55 Normally graded pyroclastic airfall–waterfall bed overlying volcaniclastic sandstone with bi-directional cross-stratification. Scale bar 15cm. *Miocene, Miura Basin, south central Japan.*

3.56 Thick sandstone bed with outsize floating blocks (two near hammer), and an oscillation grading through bed, from granule to coarse-sand sizes; part of deep-water turbidite-debrite succession.
Hammer 45cm.
Miocene, Tabernas Basin, SE Spain.

3.57 Sandstone to pebbly sandstone showing clear oscillation grading (coarsening to fining upwards, as shown by arrows) in fluvial to shallow-marine succession.
Width of view 80cm.
Permian, Bondi Beach, Sydney, Australia.

3.58 Volcaniclastic graded turbidite (arrow) showing complete Bouma sequence of structures (as labeled A–E) over sharp erosive base, and passing up into bioturbated hemi-pelagite (H). Lens cap 6.5cm.
Miocene, Miura Basin, south central Japan.

3.59 Medium-bedded sand-stone turbidite showing normal grading (arrow) and Bouma divisions ABC as marked. Base with clear scours (S), loads (L) and flame (F) structures. Knife 12cm.
Paleogene, S California, USA.

3.60 Calcarenite graded turbidites (arrows) with partial Bouma sequences (as labelled). Note that this limestone unit (mainly intrasparitic packstone and grainstone, see Chapter 12) shows very similar sedimentary structures to those of siliciclastic and volcaniclastic turbidites. Width of view 25cm.
Cretaceous, Monte Pietralata, central Italy.

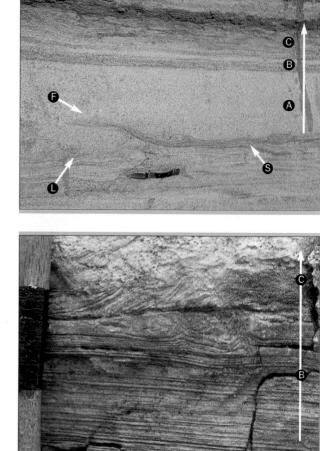

3.61 Normally graded sandstone to mudstone turbidites (arrows), medium bedded. Note sharp-based, well-preserved sandstones, but friable, poorly preserved mudstones, typical of many turbidite successions. Note also amalgamation surface (A, dashed line) in lower thick sandstone.
Photo by Paul Potter.
Cretaceous, San Luis Potosi, Mexico.

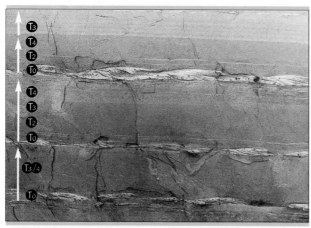

3.62 Silt-laminated mudstone turbidites showing fading ripple structures (T0) at base of fine-grained turbidite sequences (partial Stow sequences as labelled); slope apron to fan fringe setting.
Width of view 12 cm.
Cambro-Ordovician, Halifax Formation, Nova Scotia, Canada.

3.63 Detail of silt-laminated mudstone turbidites (arrows) with partial Stow sequences (as labelled), with thin parallel laminae and long-wavelength low-amplitude micro-ripples (T2 near base); lacustrine setting with volcaniclastic input. Width of view 6cm.
Triassic, Puquen, west central Chile.

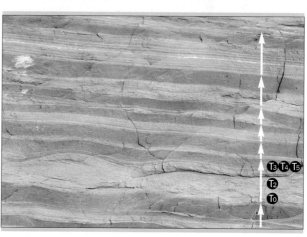

3.64 Part of very thick succession of reddish-coloured, silt-laminated mudstone turbidites (arrows) showing partial Stow sequences (T divisions as marked for one unit). Note that for the thicker siltstone portions (paler coloured) of fine-grained turbidites, the Stow T0 division equates to the Bouma C division. Width of view 20cm.
Late Precambrian, Hallett Cove, S Australia.

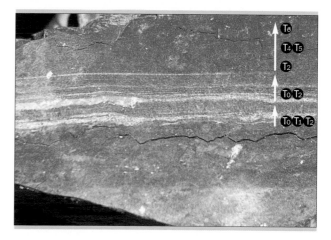

3.65 Thin-bedded, graded (arrow), silt-laminated turbidite showing Stow sequences (T divisions as marked). Width of view 20cm.
Cretaceous, central Honshu, Japan.

Post-depositional deformation and dewatering structures

THERE ARE many structures that form soon after or immediately following deposition, while the sediment is still unconsolidated or only partly consolidated, by the process of soft-sediment or wet-sediment deformation. These processes range in scale from the mass displacement of some 500km³ of sediment (slumping and sliding) to the minor shrinkage of a thin surface layer of limited extent (desiccation cracks).

Slumps and slides

(Fig. 3.14; Plates 3.66–3.74, 15.14, 15.15, 15.63)

Common to all slopes is the mass displacement of unstable sediment due to the action of some external trigger (e.g. earthquake, excess load, erosive undercutting). The scale of movement varies from cm to km.

- *Slides* show little internal deformation over a basal slip zone.
- *Slumps* show internal folding (recumbent and asymmetric) and associated thrusting.

Observe and measure:

1. Distinguish trace slump/slide from tectonic deformation and *in situ* disturbance (e.g. convolution) – note basal slip zone with shear, undisturbed beds above and below zone of disharmonic folds.
2. Measure thickness of slump/slide and, where possible, downslope and across slope dimensions.
3. Note orientation of fold axes and direction of overturned folds to ascertain direction of slumping and paleoslope.

Chaotica

(Fig. 3.14; Plates 3.74–3.81, 3.72, 4.12–4.18, 14.6–14.26, 14.18, 14.26, 15.39)

Chaotic, disorganized or structureless units, zones, or beds can be either depositional in origin (e.g. debrites, mudflows, tillites, avalanche and flood deposits – see page 61, *Structureless Beds*) or post-depositional. The most commonly encountered chaotic beds are debrites, ranging from 50cm to more than 50m in thickness. These are the deposits of debris flows, gravity driven mass movement events that occur in both subaerial and subaqueous slope environments. They comprise

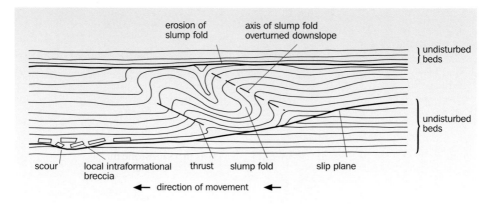

3.14 **Principal features of a slide–slump unit. Sliding involves simple lateral translation of a sediment mass along a basal dislocation zone–slip plane. Slumping involves both lateral translation and internal disruption of bedding. Both can occur at all scales from cm to km.** *Modified from Tucker (1996).*

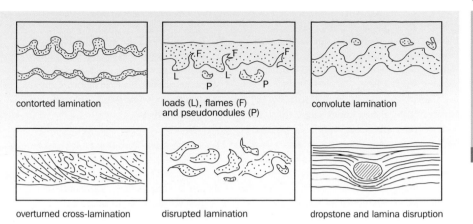

contorted lamination

loads (L), flames (F)
and pseudonodules (P)

convolute lamination

overturned cross-lamination

disrupted lamination

dropstone and lamina disruption

3.15 **Small-scale sedimentary structures that result from bedding/lamination disruption.**

a very poorly sorted mixture of clast types and sizes, including outsize clasts in very thick beds, but generally show an intrabasinal clast origin. Debrites can be entirely structureless (chaotic), show organization as part of a tripartite megabed (slide–debrite–turbidite), or show some degree of internal organization (ie crude structural divisions and clast alignment). Debris avalanche, debris slide, and mudflow deposits are all variations arising from the same basic process. Near-vent volcaniclastic fall and flow deposits, and those involving violent interaction with water, yield volcaniclastic chaotica. Glacial tillites, though very different in origin, can appear almost identical in the field to the truly chaotic debrites. Other features of glacial origin, such as striated boulders and pavements, must be found before a firm identification is made. Olistostrome is a term used more or less synonymously with very large-scale debrites, although it is has also been used (incorrectly) to describe a tectonic melange (see below) or other very large-scale slope collapse features.

Post-depositional chaotica include:

- *Fault breccias* and ground or mylonitized rock form in zones of tectonic movement.

- *Crack/fissure breccias* form by the infilling of cracks and cavities that have opened up in lithified or partly lithified rocks (especially limestones).
- *Solution/collapse breccias* form from the infilling of larger solution cavities – often surface and cave karstic hollows in limestones and evaporites.
- *Sediment injection* occurs as the result of over-pressurization of buried sediment (mainly sand and mud) and its escape into a fracture network in the surrounding rock – e.g. sandstone sills, dykes and volcanoes, mud diapirs and mud volcanoes, and salt diapirs.
- *Magma injection* into and over wet sediment results in a chaotic mix of (glassy) volcanic fragments and sediment – e.g. peperite.
- *Shale clasts* (see *Deformed Bedding*, below) can occur as chaotically arranged clasts in a sandstone matrix.
- *Melanges* (see below).

Melanges are a chaotic mixture of rock types, commonly including exotic and outsize clasts, in a finer-grained matrix. They occur as very thick bodies having structural relationships and bedding contacts that imply some element of tectonic control during emplacement (ie associated with large-scale mud volcanoes, ophiolite emplacement, major thrust faulting, etc). They also possess a distinctive tectonic shear fabric.

The scale at which such features occur is very variable. Faults zones range from a few mm to tens and even hundreds of metres. Sediment injection can be as thin pseudo-beds to irregular zones over 1km^2 in area. Melanges can cover tens of square kilometers. At the smaller (bed) scale, complete laminae disruption can result in a thin chaotic horizon, whereas shale clast chaotic beds can be several metres thick.

Observe and measure:

1. The nature and geometry of the chaotic zone and its orientation.

> **NOTE**
> Chaotica are an intriguing and still poorly understood class of sedimentary rocks. They require much careful observation in the field before one is able to fully understand and correctly interpret them.

2. The composition of material (clasts, matrix, exotic or intrabasinal).
3. Clast fabric – e.g. random or preferred orientation, jigsaw arrangement.
4. Clast interaction – e.g. fusing or partial melting, as in peperites.
5. Any signature of either glacial or tectonic influence.

Slides, slumps, and chaotica

3.66 Slump unit 2.5m thick between undisturbed beds (U) above and below. Basal slip plane (solid line with arrow), internal thrust (dashed line), and slump fold (S) indicate sense of downslope movement from left to right.
Cretaceous, Umbro-Marche, central Italy

3.67 Small-scale slide–slump unit comprising mudstone–sandstone (partly carbonate cemented) over pro-delta silt-stones and sandstones. Slip plane dashed, sense of movement to right.
Notebook 15cm high.
Cretaceous, central California, USA.

3.68 Part of relatively thin (4–5m) and laterally extensive (approximately 25km²) slide-slump unit in carbonate slope succession. Undisturbed beds (U) at top and base, basal slip plane (solid line), zone with internal thrust planes (dashed lines) passes laterally into more chaotic section (right).
Height of section 10m.
Miocene, southern Cyprus.

3.69 Detail of contorted strata in slide–slump unit.
Hammer 45cm.
Miocene, southern Cyprus.

3.70 Chilean geologist (Manuel Suarez) standing in nose of slump fold within turbidite slope-apron succession.
Triassic, Los Molles, west central Chile.

3.71 Overturned and dislocated limb of slump fold in turbidite slope-apron succession. Width of view 6m.
Oligo–Miocene, N Sicily, Italy.

3.72 Slope-apron succession adjacent to carbonate shelf with large slump-disrupted and slide-stacked shelf edge limestone–marl units within chaotic debrite matrix. Height of cliff approximately 100m.
Oligocene, El Charco, SE Spain

3.73 Small-scale slump, showing contorted silt-laminated turbiditic mudstone unit. Normal bedding just in view at base of picture, and just out of view at top.
Width of view 12cm.
Paleogene, central California, USA.

3.74 Nose of slump-folded turbidite succession with internal dislocation thrust (dashed line); note that this unit is interpreted as forming a single large clast within a debrite megabed (Gordo Megabed).
Miocene, Tabernas Basin, SE Spain.

3.75 Fault breccia, sub-vertical, at margin of Jurassic-age limestone succession. Hammer 45cm. *SE Cephallonia, Greece.*

3.76 Ice-wedge fissure breccia fill through lacustrine mudstone succession. Width of view 1m. *Pleistocene, Rocky Mountains, Alberta, Canada.*

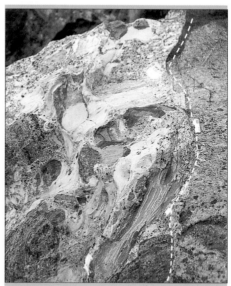

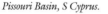

3.77 Solution collapse breccia of limestone blocks and overlying alluvial gravels into karstic cavity or solution hollow in underlying Miocene limestone succession. Karst surface marked with dashed line. Width of view 5m. *Pissouri Basin, S Cyprus.*

3.78 Injectionite, comprising igneous conglomerate (breccia), formed by diapiric injection of mixed igneous and sedimentary material (left of dashed line) into forearc slope-apron succession (right). Width of view 2m. *Miocene, Miura Basin, south central Japan.*

3.79 Sandstone injection, from deep-water massive sandstone unit, through slope mudstones. Width of view 15m.
Oligo–Miocene, Numidian Flysch, N Sicily, Italy.

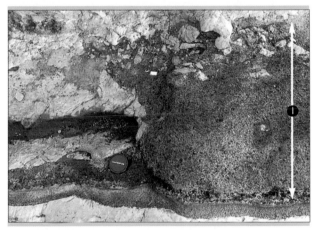

3.80 Detail of injectionite unit (I) of dark volcaniclastic sandstone mixed with disrupted pale hemipelagic mudstones, within vocaniclastic forearc basin succession. Undisturbed beds left of centre. Width of view 50cm.
Miocene, Miura Basin, S Central Japan.

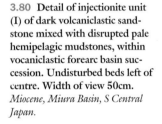

3.81 Part of very large scale melange complex (tens of km²) associated with emplacement of the Troodos ophiolite. Blocks include: Triassic dolomitized limestone (part of reef talus breccia, R), Cretaceous seafloor pillow basalts (B), Paleogene deep-sea micrites (M). The matrix is a weakly sheared shale (S) with no visible bedding.
Age uncertain, Aphrodites Bay, S Cyprus.

Deformed bedding and shale clasts

(*Fig. 3.15–3.17; Plates* 3.82–3.89, 5.17, 6.6, 6.7, 6.22)

At the small and medium scale there are a range of post-depositional processes that operate to disturb or deform the primary structures more or less *in situ* – i.e. without significant lateral displacement. These include the following:

- *Convolute lamination* results from flow-induced shear and frictional drag on incipient ripples – asymmetric and over-turned in flow direction. This occurs as the T1 division of the Stow sequence, and may occur in addition to or in place of ripple cross-lamination (C division) of the Bouma sequence of turbidites.
- *Contorted lamination* is less regular and with no preferred orientation and can result from seismic shock, liquefaction, and sediment dewatering.
- *Disrupted lamination* refers to the more complete deformation/brecciation of laminae or beds – in some cases developed from convolute or contorted lamination by very rapid dumping of sediment load. This can sometimes be confused with extensive bioturbation.
- *Overturned cross-bedding* forms as over-steepened ripple/dune foresets collapse in a downflow direction, generally as a result of over-rapid deposition from high-energy, sediment charged flows.
- *Shale clasts* (also rip-up clasts) form from more intensive bed disruption and partial or complete erosion of the underlying bed, through the passage of a strongly erosive current, and also from bank collapse into a passing flow. A wide range of types and sizes occur. Softer sediment may yield mud clasts, fine-grained limestones yield micrite clasts, and so on.
- *Loads, flames and pseudonodules* form as a result of differential sinking of one bed into another – typically sand into mud.

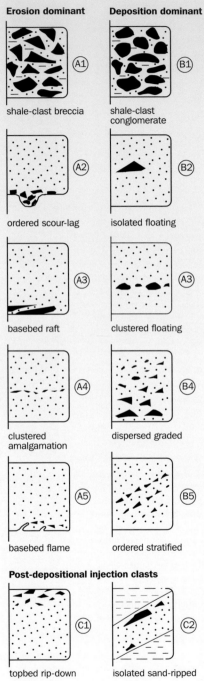

Erosion dominant   **Deposition dominant**

(A1) shale-clast breccia   (B1) shale-clast conglomerate

(A2) ordered scour-lag   (B2) isolated floating

(A3) basebed raft   (A3) clustered floating

(A4) clustered amalgamation   (B4) dispersed graded

(A5) basebed flame   (B5) ordered stratified

Post-depositional injection clasts

(C1) topbed rip-down   (C2) isolated sand-ripped

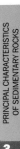

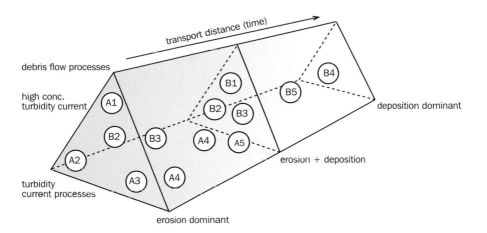

3.17 Interpretation of shale clast types (as numbered in *Fig 3.15*) in terms of depositional process and transport distance. Those listed as 'erosion dominant' have only recently been incorporated into the flow; 'deposition dominant' types have been shaped and positioned by longer duration of the flow process.

- ***Dropstones*** that fall from floating ice (also from seaweed and so on) as well as volcanic bombs/ejecta may fall onto a soft sediment surface causing local depression of the laminae around the dropstone.
- ***Raindrop imprints*** cause very minor disturbance to the bed surface, but leave a very distinctive pattern of minor depressions and rims.

Observe and measure:
1. The nature and scale of deformed bedding, where it occurs within the bed, and if it forms part of any structural sequence.
2. Any evidence of flow direction.
3. The type, nature, abundance, and orientation of shale clasts.
4. Evidence of way-up of strata.

3.16 The principal types of shale clast that occur in deep-water massive sands and associated deposits. *Modified from Johansson & Stow (1995), Geol. Soc. Special Publ. 94, 211–241.*

Water-escape and desiccation structures
(*Fig. 3.18, 3.19; Plates* **3.73–3.80**)

A variety of deformational structures result from the lateral and upward passage of water through sediment. This can occur rapidly immediately after deposition (especially on sudden dumping from turbulent suspension) or more slowly through time. Structures include the following:

- ***Dish structures*** of concave-up laminae – cm to dm scale.
- ***Consolidation laminae*** indicating more extensive lateral movement of water.
- ***Sheet and pipe*** (also pillar) structures caused by the upward flow of water.
- ***Sand volcanoes*** formed at the bed surface where fluid and sand has escaped.
- ***Burst-through structures*** formed by small-scale fluid breakthrough of laminated sediment.
- ***Contorted lamination*** formed as a result of water escape disrupting the lamination (see *Deformed Bedding*, above).
- ***Syneresis cracks*** formed through the slow dewatering of seafloor or lake floor sediment – trilete and spindle-shaped forms.

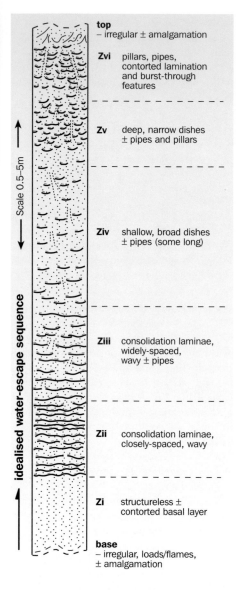

Scale 0.5–5m

idealised water-escape sequence

top
– irregular ± amalgamation

Zvi pillars, pipes, contorted lamination and burst-through features

Zv deep, narrow dishes ± pipes and pillars

Ziv shallow, broad dishes ± pipes (some long)

Ziii consolidation laminae, widely-spaced, wavy ± pipes

Zii consolidation laminae, closely-spaced, wavy

Zi structureless ± contorted basal layer

base
– irregular, loads/flames, ± amalgamation

3.18 Idealized vertical sequence of water escape structures, typical in thick sand-rich turbidites, deep-water massive sands, and associated deposits. The scale over which this sequence typically occurs is 0.5–5m.
Modified from Stow & Johansson (2000), Marine & Petroleum Geology 17, 145–174.

Continued from previous page...

- ***Desiccation cracks*** (mudcracks) formed through more complete drying up of the surface layers of (muddy) sediment on sub-aerial exposure leading to shrinkage, cracking and infill – typically of polygonal form, centimetre–metre scale.

Observe and measure:
1. The nature and scale of water-escape/ desiccation structures, including dish wavelength.
2. The thickness and lateral extent of bed(s) affected.
3. Any vertical organization of water-escape structures.

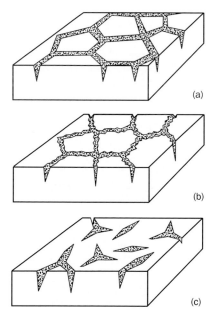

(a)

(b)

(c)

3.19 Shrinkage cracks formed by subaerial desiccation (a, b), and syneresis cracks formed by shallow subaqueous water escape (c). Both types vary in size, in regularity of shape and in depth of penetration into the underlying sediment (up to several tens of cm). The filled V-shaped cracks in cross section may become ptygmatically folded due to compaction.

Deformed bedding and shale clasts

3.82 Thin horizon showing load-induced convolute lamination in volcaniclastic siltstone–mudstone turbidite sandwiched between thick-bedded sandstone turbidites. Width of view 15cm. *Pliocene, Boso Peninsula, near Tokyo, Japan.*

3.83 Highly disorganized and contorted siltstone–mudstone units within turbidite slope succession; interpreted as very rapid dumping of load from turbidity current as it spills over channel levee. Note that this is not bioturbation. Hammer 25cm. *Triassic–Jurassic, Los Molles, west central Chile.*

3.84 Loading and possible scouring of siltstone/sandstone into mudstone, within interbedded siltstone/sandstone turbidites and dark grey mudstone turbidites and hemipelagites. Hammer 30cm. *Eocene, near Annot, SE France.*

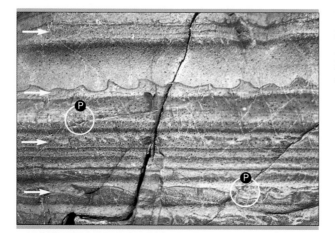

3.85 Load (L) and flame (F) structures at base of normally graded volcaniclastic turbidite. Pale mud floating clasts within turbidite probably derived from detachment of flame structures. *Miocene, Miura Basin, south central Japan.*

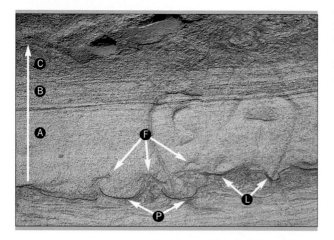

3.86 Loads, flames and pseudonodules – P (as marked) within volcaniclastic turbidite succession. Width of view 25cm. Photo by Bob Foster. *Muzwezwe River, Zimbabwe.*

3.87 Loads (L), flames (F), and pseudonodules (P) along base of graded sandstone turbidite (Bouma divisions ABC as marked). Width of view 30cm. *Paleogene, S California, USA.*

3.88 Shale-clast horizon (arrow) within structureless sandstone, representing broken-up mudstone interval between sandy turbidites.
Width of view 40cm.
Eocene, Pera Cava, SE France.

3.89 Shale-clast-rich turbidite sandstone within deep-water turbidite succession; clasts possibly derived from local channel-bank collapse. The many different types of shale clast, their origins and occurrence are illustrated in *Figs 3.16, 3.17*. Hammer 25cm.
Cretaceous, Carmelo, central California, USA.

Water-escape and desiccation structures

3.90 Water-escape burst-through structures (B) deforming parallel and cross-lamination into convolute lamination; deep-water turbidite succession. Width of view 30cm. *Oligocene, Reitano Flysch, NE Sicily, Italy*.

3.91 Large tepee structure in peritidal limestones, formed on the lee side of a barrier island. Tepee structures range from small-scale buckled polygonal desiccation cracks (typical desiccation structures on carbonate tidal flats) to the larger and more complex features pictured here. Photo by Paul Potter. Width of view 2.5m. *Permian, Carlsbad Cavern National Park, W Texas, USA*.

3.92 Water-escape dish structures, together with burst-through/short pipe structures, in deep-water massive (turbidite) sandstone succession. Width of view 40cm. *Eocene, Cantua Basin, central California, USA*.

3.93 Water-escape dish structures (detail) in deep-water massive (turbidite) sandstone succession. Width of view 20cm. *Eocene, Cantua Basin, central California, USA.*

3.94 Water-escape pipe and sheet structures in deep-water massive (turbidite) sandstones. Width of view 30cm. *Oligo–Miocene, Numidian Flysch, N Sicily, Italy.*

3.95 Convolute lamination and burst-through structure (B) at top of sandstone turbidite bed. Lens cap 6cm. *Paleogene, southern California, USA.*

3.96 Vertical pipe/chimney structure indicative of large-scale water escape, in fan-delta sandstone succession. Hammer 45cm.
Pliocene, near Carboneras, SE Spain.

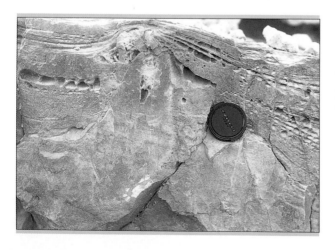

3.97 Vent of small sand volcano (left of lens cap) through calcarenite turbidite, now carbonate cemented and capped with parallel-laminated sandstone (upper division of turbidite). Lens cap 6cm.
Cretaceous, Ifach, SE Spain.

3.98 Bedding plane view of syneresis cracks (trilete and irregular star-shaped) on surface of mudstone bed. These form by subaqueous dewatering of sediments.
Jurassic, near Whitby, NE England.

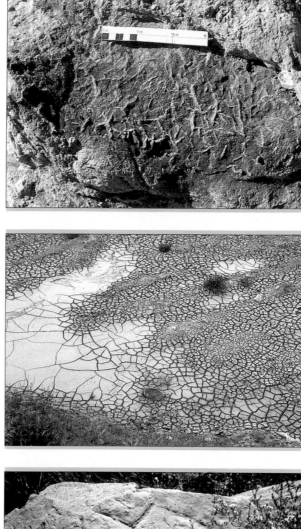

3.99 Present day mudcracks (desiccation cracks) on dried out surface of raised mudflat.
Recent, East Anglia, E England.

3.100 Open irregular polygonal mudcracks (desiccation cracks) formed on ancient supratidal carbonate mudflats.
Photo by Paul Potter.
Hammer 25cm.
Ordovician, Burksville, Cumberland, Kentucky, USA.

Biogenic sedimentary structures

THERE ARE MANY structures formed in sediments by the action of plants and animals. These include irregular disruption of the sediments (bioturbation), discrete organized markings (trace fossils or ichnofossils), and biogenic growth structures (e.g. stromatolites). Some modern organic markings (such as borings and surface trails on rocks), as well as inorganic structures (such as syneresis cracks, water-escape structures, concretions) can be confused with trace fossils in certain instances.

Bioturbation
(*Fig. 3.20; Plates* 3.101–3.120)

Bioturbation refers to the irregular disruption of sediment by plants and animals, rather than organized and recognizable burrows or other traces. It is commonly described in terms of intensity (or percentage bioturbated), ranging from no or sparse bioturbation to intense or complete bioturbation. The sediment appearance changes from slightly mottled, often with distinct trace fossils, to thoroughly churned. Bioturbation always accompanies trace fossil activity.

Observe and measure:
1. The degree or intensity of bioturbation.
2. The different sediment types intermixed.
3. Any distinct trace fossils (see below).

Trace fossils
(*Fig. 3.21, 3.22, 3.23; Plates* 3.101–3.122)

The study of trace fossils (ichnofossils) as part of sediment facies can reveal much complementary information to that gained from observing primary (dynamic) structures on the one hand and true body fossils on the other. Although they are treated in some respects as any other type of fossil, having ichnogenus and ichnospecies names, they are formed very much by the interaction of organisms with the sediment and only very rarely reveal the true identity of their architects (ie the organism that formed them).

Trace fossils can be classified in terms of their mode of preservation, within a mudstone or sandstone bed or at a boundary between the two, but this does little more than convey information regarding their morphological expression (toponomy). A more revealing classification, preferred here, is in terms of the behaviour (ethology) of the organism that formed them. This, after all, provides greater insight into the depositional environment. There are now 13 different groups recognized in this classification, necessarily with some behavioural overlap between groups.

Grade	Percent	Classification bioturbated
0	0	No bioturbation
1	1–4	Sparse bioturbation, bedding distinct, few discrete traces and/or escape structures
2	5–30	Low bioturbation, bedding distinct, low trace density, escape structures may be common
3	31–60	Moderate bioturbation, bedding boundaries sharp, traces discrete, overlap rare
4	61–90	High bioturbation, bedding boundaries indistinct, high trace density with overlap common
5	91–99	Intense bioturbation, bedding completely disturbed (but just visible), limited reworking, later burrows discrete
6	100	Complete bioturbation, sediment reworking due to repeated overprinting

3.20 **Bioturbation index** (*after Tucker 1996*).
An attempt to quantify or grade bioturbation in terms of the degree of primary fabric remaining, the abundance of burrows and the amount of burrow overlap.

resting + crawling traces
e.g. *Rusophycos* & *Cruziana*

surface track/trail
e.g. *Chirotherium*

resting trace e.g. *Asteriacites*

grazing trace e.g. *Nereites*

regular burrow network
grazing trace
e.g. *Paleodictyon*

spirally coiled grazing trace
e.g. *Spirorhaphe*

simple straight tubes
eg. *Skolithos*

branching dwelling burrow
with pelleted walls
e.g. *Ophiomorpha*

organised feeding burrow
system e.g. *Chondrites*

vertical U-tube with spreite
e.g. *Diplocraterion*

branching dwelling burrow
e.g. *Thalassinoides*

horizontal feeding burrow
system e.g. *Zoophycos*

horizontal – subhorizontal
U-tube with spreite
e.g. *Rhizocorallium*

boring traces into hard substrate
e.g. *Trypanites*

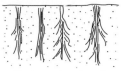

rootlet traces in paleosol

3.21 Examples of trace fossils, showing type of behaviour by organism responsible for the trace, and one or more examples of ichnofossil genus. For examples and scales see *Photos* 3.101–3.122.

The main trace fossil groups recognized are as follows:

- *Crawling traces* (repichnia) are relatively simple, linear to sinuous trails and tracks made by animals on the move over the surface of the sediment – e.g. bird/dinosaur footprints, crustacean/trilobite tracks, as well as true crawling marks of various types of annelid.
- *Grazing traces* (pascichnia) are more complicated trails (meandering, coiled, radiating) produced by deposit-feeders systematically working over the surface sediment for food.
- *Traps and gardening traces* (agrichnia) are regular patterned structures, related to grazing taces but still more systematic in form.
- *Resting traces* (cubichnia) are the body impressions of animals temporarily at rest on or just below the sediment surface – e.g. starfish and bivalve impressions.
- *Dwelling traces* (domichnia) are simple to complex, horizontal, inclined or vertical, single or U-shaped burrows in which the animal lived – they can have lined or pelleted walls, and concave-up laminae (spreite) between the U-tubes.
- *Feeding traces* (fodichnia) are simple to complex, branched or unbranched, well-organized burrow systems formed by deposit-feeders searching systematically through the sediment for food; they are often also used for dwelling purposes.
- *Boring traces* are tubular to rounded holes, generally infilled with sediment, made by specific bivalves, sponges and annelids into early cemented seafloor (hardgrounds), pebbles, shale clasts and larger fossils. These are grouped with other predation traces (praedichnia).
- *Rootlet traces* are simple to complex, branched or unbranched, mainly vertical to inclined, but also horizontal systems formed by plant roots penetrating sediment – typically carbonized with black streaks remaining, also the site for concretions (rhizocretions).
- *Other types* of trace fossil are formed by animals: attempting to maintain a depth equilibrium beneath gradually aggrading or degrading sea floors (equilibrichnia); trying to escape through the sediment following sudden burial (fugichnia); building constructions largely of sediment for dwelling (aedificichnia), or for breeding (calichnia) purposes.

Trace fossil analysis yields information on:
- Depositional environment and water depth (the Seilacher ichnofacies concept, *Fig. 3.22*).
- Rates and styles of deposition, from the abundance, diversity and tiering structure observed (*Fig.3.23*).
- Any limiting stress factors such as oxygen abundance or salinity levels (fewer, smaller traces and reduced diversity at higher stress levels).
- Sequence stratigraphic markers of environmental change (e.g. *Fig 3.23*).

Observe and measure:
1. The different tracks, trails, and impressions on bed surfaces; burrows, borings, and rootlet traces within beds.
2. The size, shape, orientation and spacing of traces. Burrows can be simple and straight, curved and regular, branching and complex, horizontal or vertical.
3. The nature of the burrow/trace wall and fill and the possible derivation of fill material (e.g. from adjacent beds). Burrows can be passively filled by sediment collapse or actively filled by lining the walls and packing fecal pellets into distinct curved menisci.
4. The presence of a horizon or hardground marking a paleosurface on and below which burrowing activity took place. Such horizons are known as omission surfaces and signify a distinct change in environment and hence ichnofacies.

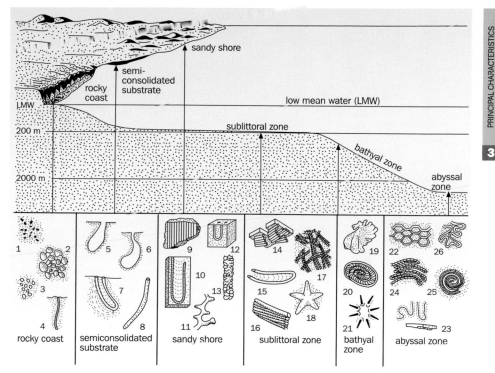

3.22 Summary of the most common trace fossils (as numbered) and ichnofacies (as named after the most common ichnofossils present). Typical environmental occurrence of each ichnofacies is also indicated. Trace fossils as follows: 1 *Caulostrepis*; 2 *Entobia*; 3 echinoid borings; 4 *Trypanites*; 5, 6 *Gastrochaenolites*; 7 *Diplocraterion*; 8 *Psilonichnus*; 9 *Skolithos*; 10 *Diplocraterion*; 11 *Thalassinoides*; 12 *Arenicolites*; 13 *Ophiomorpha*; 14 *Phycodes*; 15 *Rhizocorallium*; 16 *Teichichnus*; 17 *Crossopodia*; 18 *Asteriacites*; 19 *Zoophycos*; 20 *Lorenzinia*; 21 *Zoophycos*; 22 *Paleodictyon*; 23 *Taphrhelminthopsis*; 24 *Helminthoida*; 25 *Spiroraphe*; 26 *Cosmoraphe*. From Frey & Pemberton, in Walker (1984).

5. The occurrence of a distinctive tiering system of different burrows at different depths relative to the original seafloor, including any cross-cutting relationships. Different organisms live and work at different levels in the substrate – the more tiers of trace fossils present, the more stable and mature the environment.

Tiering
(*Fig. 3.23; Plates* **3.106, 3.114, 3.116, 3.119**)

Different organisms occupy different depths in the substrate. Some crawl or graze on the

sediment surface, others rest or live just below it, while still others construct much deeper burrows for dwelling or feeding. Most burrowing occurs within 50cm below the surface, though deeper traces can be found. Within this trace-fossil zone, the presence of two or more tiers, characterized by particular traces or trace assemblages, reveals the progressive colonisation of a virgin substrate by an influx of organisms. Certain opportunistic forms arrive first, others more gradually, but may then dig deeper and work the sediment more thoroughly. The more tiers present, the more stable and mature the environment.

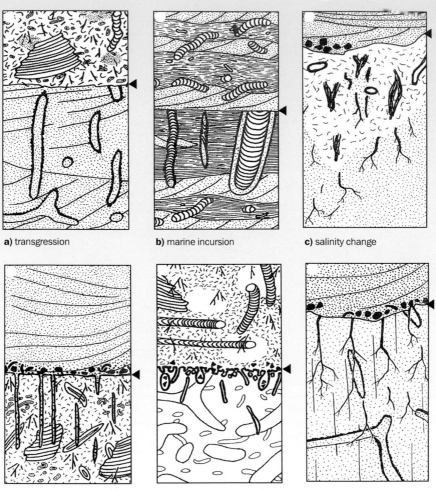

a) transgression   **b)** marine incursion   **c)** salinity change

d) firmground omission surface   **e)** hardground omission surface   **f)** rooted shoreface

3.23 Trace fossil tiering related to key omission and other stratal surfaces marked by arrows (*Modified from Bromley 1990*).
a) Shoreface sands with *Ophiomorpha* overlain by bioturbated offshore mud with *Chondrites*, *Teichichnus*, *Phycosiphon*, and *Planolites*.
b) Non-marine mixed facies with *Taenidium* traces cut by short marine incursion yielding *Diplocraterion* and *Skolithos*.
c) Non-marine facies with rootlet system cut by marine incursion bringing *Roselia*, *Planolites*, and *Paleophycos* traces.

d) Omission surface over muddy offshore facies with mixed ichnofacies allows hardground development and opportunistic *Skolithos* traces. Shallow marine sands above.
e) *Thallassinoides* and *Planolites* in softer sediments below cemented omission surface with boring traces. Overlain by bioturbated offshore mud with *Chondrites*, *Zoophycos*, *Teichichnus*, and *Planolites*.
f) Fluvial sand over shoreface sands. Rootlet traces now penetrating earlier formed *Ophiomorpha* and *Skolithos* traces.

Biogenic growth structures (stromatolites)

(Fig. 3.24; Plates 3.123, 7.11, 7.14–7.20, 15.52–15.54)

Some limestones are created directly through the action of organisms that construct their own firm habitats by the secretion of calcium carbonate. These include reefs and bioherms formed by corals, bryozoans, sponges, algae, and molluscs (see Chapter 7). Another distinctive type of crinkly lamination in limestone sediments is formed as a result of the trapping and binding of carbonate particles by a surficial microbial mat (formerly algal mat), mainly composed of cyanobacteria. The carbonate particles consist of (biochemical) micrite, peloids and skeletal debris. These growth structures, known as stromatolites, have a great variety of forms:

- Planar microbial laminites ('algal' laminites) have crinkly or corrugated lamination, commonly disrupted by desiccation cracks, elongate cavities (laminar fenestrae), and local grainstone horizons.
- Domal stromatolites have crinkly lamination semi-continuous between individual domes.
- Columnar stromatolites are discrete, very partially linked, single or branched columnar structures with intraclasts and carbonate grains between columns.
- Oncolites (or oncoids) are subspherical, unattached, microbial laminated structures, with crude concentric lamination that is typically asymmetric and discontinuous.

Observe and measure:

1. The nature of lamination and the three-dimensional form of growth structure.
2. Any distinctive evolution from base to top of the stromalotic zone, commensurate with changing depositional environment.

planar microbial laminites with fenestrae and desiccation cracks

crinkly microbial laminites (± features a/a)

domal stomatolites with microbial laminites continuous between domes

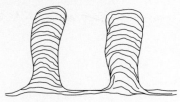

columnar stromatolites ± various forms, simple or branching, minor projections, margin structures

oncolites, sub-spherical form with nucleus and symmetric/asymmetric microbial laminar growth

3.24 **Microbial laminites, stromatolites and oncolites.**

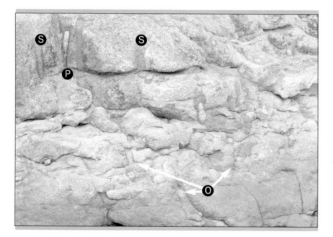

Bioturbation and biogenic structures

3.101 Simple vertical dwelling burrows, *Skolithos* (S), intertidal to shallow marine. Width of view 30cm. *Jurassic, Osmington Mills, S England.*

3.102 Vertical dwelling burrows, including *Skolithos* (S) and the Y-shaped *Polychladichnus* (P). Opportunistic fauna colonizing newly established substrate after environmental change. Lower part of view has T-branched boxwork of burrows with thick, externally knobbly linings. These are *Ophiomorpha* (O). Width of view 30cm. *Jurassic, Osmington Mills, S England.*

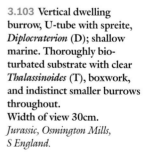

3.103 Vertical dwelling burrow, U-tube with spreite, *Diplocraterion* (D); shallow marine. Thoroughly bioturbated substrate with clear *Thalassinoides* (T), boxwork, and indistinct smaller burrows throughout. Width of view 30cm. *Jurassic, Osmington Mills, S England.*

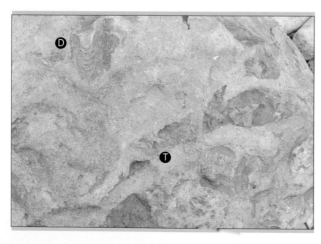

3.104 Network of vertical and subvertical burrows representing the deposit-feeding and dwelling trace *Macaronichus* (M). Occurs throughout this sandy prodelta section, crosscutting the large-scale foreset lamination, slightly inclined to left; shallow marine.
Lens cap 6cm
Pliocene, Pissouri, S Cyprus.

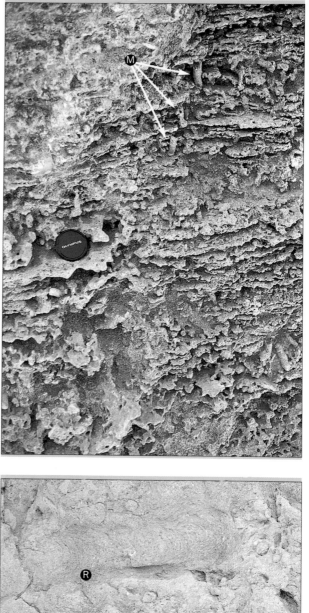

3.105 Sub-horizontal feeding/dwelling burrows, U-tube with spreite, *Rhizocorallium* (R); shallow marine. Several traces and ghost traces are visible in this view. Width of view 30cm.
Jurassic, Osmington Mills, S England.

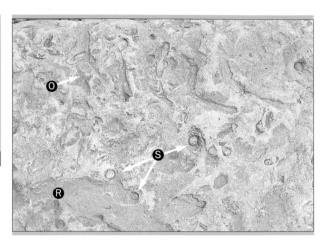

3.106 Mixed assemblage, bedding-plane view, including *Rhizocorallium* (R), *Ophiomorpha* (O), and cross-sections of *Skolithos* (S). Other burrows and general bioturbation present but not distinct. Width of view 30cm.
Jurassic, Osmington Mills, S England.

3.107 Probable crustacean burrow network, *Macaronichus*, within shallow intertidal muddy sandstones. Width of view 20cm.
Triassic, Los Molles, west central Chile.

3.108 Complex dwelling burrow network, *Thalassinoides*; shallow marine. Width of view 20cm.
Miocene, Sorbas, SE Spain.

3.109 Endobenthic complex
feeding burrow (looking down
on part of surficial mound),
Zoophycos (Z), common from
shelf to deep water.
Lens cap 6cm.
*Cretaceous, Scaglia Rossa, central
Italy.*

3.110 Bedding-plane view of
typical, faint *Zoophycos* traces,
deep-water micrites.
Width of view 20cm.
*Paleogene, Petra Tou Romiou,
S Cyprus.*

3.111 Endobenthic complex
feeding burrow, *Chondrites*,
common from shelf to deep
water. **Lens cap 6cm.**
*Miocene, Miura Basin, south
central Japan.*

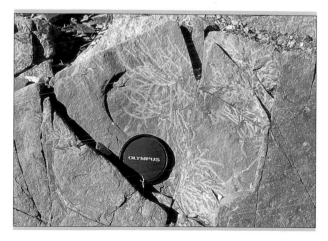

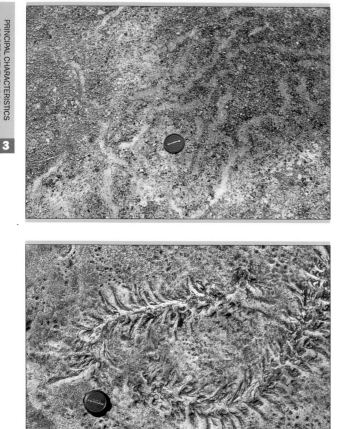

3.112 Horizontal surface grazing trace, uncertain ichnogenus. Bedding-plane view of hemipelagic sediment within deep-water forearc basin. Width of view 80cm.
Pliocene, Boso Peninsula, south central Japan.

3.113 Surface crawling trace, *Nereites*, common in deeper water. Bedding-plane view of surface of fine-grained turbidite. Width of view 60cm
Miocene, Monterey Formation, California, USA.

3.114 Dense population of *Diplocraterion* (U-shaped with spreite) over irregular, nodular horizon (lower view) created by large *Thalassinoides* burrows. Width of view 20cm.
Jurassic, Osmington Mills, S England.

3.115 Probable dwelling traces, *Paleophycos* (few marked P), together with mixed assemblage, including indistinct *Thalassinoides* and *Planolites* burrows. Width of view 25cm. *Jurassic, Osmington Mills, S England.*

3.116 Shallow water assemblage dominated by *Planolites* (few marked P), together with other less distinct traces. Width of view 25cm. *Neogene, Yermasoyia, S. Cyprus.*

3.117 Mixed burrow assemblage with intense bioturbation; deep water. Some *Chondrites* (C) and *Paleophytes* (P); much of the less distinct burrowing is probably *Planolites*. Width of view 20cm. *Miocene, Miura Basin, south central Japan.*

3.118 Bird footprint on lagoonal mudflat; part of a longer trail of footprints – *repichnia*. Width of view 5cm.
Pleistocene, Lake Bonneville, Colorado Canyon, USA

3.119 Bioturbation obscuring parts of mixed burrow assemblage; deep-water succession. Lens cap 6cm.
Miocene, Miura Basin, south central Japan.

3.120 Intense bioturbation without clear burrow traces; shallow-water succession. Hammer 40cm.
Miocene, Pohang Basin, SE Korea.

3.121 Calcretized rootlet traces (zone **R**) preserved near the hardened (calcrete) surface of a modern eolian dune complex. Width of view 60cm.
Recent/sub-Recent, Sahara Desert, southern Tunisia.

3.122 Carbonized rootlet traces preserved in sandstone below former peat layer or paleosol horizon. Lens cap 6cm.
Jurassic, near Whitby, NE England.

3.123 Micritic limestone with oncolites (microbial growth structure); note also brown iron-staining from iron substitutions in the carbonate lattice. Two examples indicated (**O**). Width of view 15cm.
Neogene, Roldan Reef complex, Carboneras, SE Spain.

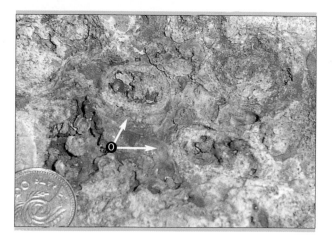

Chemogenic sedimentary structures

(*Plates* 3.124–3.138)

FROM THE POINT of deposition, through progressive burial, and as a result of uplift and exposure, the physico–chemical conditions within a sedimentary succession lead to a variety of chemogenic structures. These may destroy, disrupt or even enhance primary features, and in some cases mimic primary structures. In all cases they yield information about the physico–chemical changes that have affected the sediment to date.

Induration and induration surfaces

(*Plates* 6.1, 6.20, 7.2, 8.12, 11.10, 15.3, 15.5)

The induration or hardness of a sedimentary rock depends on original lithology, age, burial depth and degree of cementation. Induration and lithology both influence the subsequent effects of exposure and weathering and hence the appearance of the rock at outcrop. Cycles and sequences in rock successions are thus often easily seen by observing the weathering profile in the field. Mudrocks and marlstones will tend to be less indurated than sandstones, conglomerates and limestones, and so be less well exposed, tending to weather in rather than stand out. A qualitative scheme for describing the induration of sedimentary rocks is as follows:

- *Unconsolidated:* Loose; no cementation.
- *Very friable:* Crumbles easily between fingers.
- *Friable:* Grains separate and rub off between fingers; gentle blow with hammer disintegrates sample.
- *Hard:* Some grains can be separated with penknife; breaks easily when struck with hammer.
- *Very hard:* Grains difficult to separate; rock difficult to break with hammer.
- *Extremely hard:* Will only break when subjected to sharp, hard hammer blow; sample breaks across grains.

In some settings, the sediment surface becomes indurated by cementation either shortly after deposition or following uplift and exposure. Such induration surfaces include the following:

- *Hardgrounds* result from synsedimentary cementation of the seafloor, typically followed by organic encrustations, boring, current erosion or seafloor dissolution, expansion and cracking to form tepee-structures (pseudo-anticlines); commonly nodular surfaces, with or without coating/impregnation by iron minerals, phosphate and glauconite; mostly in limestones, more rarely as ferromanganese nodules/pavement in deep-sea mudrocks (*Plates* 3.124 7.1, 12.17, 12.18).
- *Paleokarstic surfaces* result from the subaerial exposure of carbonates followed by weathering and dissolution under the influence of meteoric waters; characterized by irregular topography, potholes, collapse breccias, thin patchy residual soils and laminated crusts from pedogenic processes (*Plate* 3.77).
- *Paleosols* and *duricrusts* (see Chapter 13) form as a result of subaerial weathering processes, including extensive leaching and precipitation (*Plates* 13.1–13.14).
- *Beachrock* is a carbonate-cemented beach deposit. Cementation of this kind can be extremely rapid (years or decades) so that human artifacts (bricks, drinks cans, etc) can be found as 'trace fossils' in beachrock.

Observe and measure:

1. The degree of induration and any distinctive weathering profile.
2. The occurrence and nature of any induration surface.
3. The thickness, extent and nature of any paleosol horizon; the composition of duricrusts.

Nodules or concretions

(*Fig. 3.25*; *Plates* 3.125–3.131, 3.137, 6.11, 6.13, 7.6, 7.22, 8.2, 8.3, 8.6–8.8, 15.49)

Nodules (also called concretions) are localized patches that show differential cementation from that of the host sediment. They may form at any time after deposition during burial, but are mainly early diagenetic in origin. The formation of pre- and post-compaction nodules is illustrated in *Fig.3.25*. They may be randomly dispersed, concentrated along particular horizons (especially bedding planes), or nucleated around local inhomogeneities (e.g. fossils, roots, burrows, and so on). Their size varies from a few mm to several metres in diameter, and shape from spherical to elongate to highly irregular.

The main types include:

- *Calcite nodules:* one of the most common types, especially in mudstones, sandstones, and soils; all sizes up to several metres (known as doggers); in soils they are also known as calcrete or cornstone nodules, and as rhizocretions where they develop around roots.
- *Dolomite nodules:* less common but similar occurrence to calcite nodules.
- *Siderite nodules:* typical of organic-rich, non-marine or brackish-water mudstones.
- *Pyrite nodules:* typical of organic-rich, marine mudstones.
- *Phosphate (collophane) nodules:* also typical of organic-rich, marine mudstones.
- *Gypsum–anhydrite nodules:* typical of mudstones and soils in evaporite successions. Can be replaced by quartz/celestite.

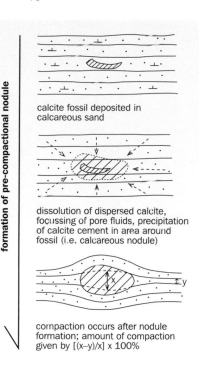

formation of pre-compactional nodule

calcite fossil deposited in calcareous sand

dissolution of dispersed calcite, focussing of pore fluids, precipitation of calcite cement in area around fossil (i.e. calcareous nodule)

compaction occurs after nodule formation; amount of compaction given by [(x–y)/x] x 100%

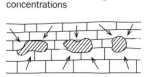

formation of post-compactional nodule

dispersed siliceous microfossils deposited in calcareous ooze

sediment compaction and dewatering ± incipient precipitation of siliceous-rich material around original concentrations

precipitation of chert nodules in chalk during late-stage pore-fluid movement

3.25 **Formation of pre- and post-compaction nodules.**

- **Quartz (chert/flint) nodules:** one of the most common types, especially in carbonate rocks. Many of the flint nodules in chalks are nucleated in burrow networks of *Thalassinoides*.
- **Septarian nodules:** these have radial and concentric cracks, formed by nodule contraction and peripheral growth, and filled with calcite or siderite crystals.
- **Geodes:** these have hollow centres with crystals of (mainly) quartz, calcite or dolomite growing towards the centre.
- **Cone-in-cone nodules:** these have fibrous calcite crystals arranged in a zig-zag cone-like pattern at right-angles to bedding, typical of some organic-rich mudrocks.

Observe and measure:
1. Nodule composition and type (as above), as well as nature of the host sediment.
2. Size, shape and distribution.
3. Any particular nucleus (e.g. fossil) or form (e.g. root burrow).
4. Any signs of early diagenetic origin (preserved fossils or burrows, deflected lamination), or late-diagenetic origin (crushed fossils, flattened burrows, undeflected lamination).

Compaction, pressure dissolution, and cavity structures
(*Fig. 3.26; Plates* **3.133**, **3.134**, **3.138**)

Progressive sediment burial leads to dewatering and compaction. The principal effects of compaction are:

- Reduction in thickness – progressive according to burial depth and cementation, up to 80–90% reduction in fine-grained sediments (muds, marls, chalk), 40–60% in coarse-grained sediments (sands, gravels, grainstones, and so on).
- Flattening of burrow traces.
- Crushing of organic remains (fossils).
- Breaking of platy mineral grains (e.g. micas).

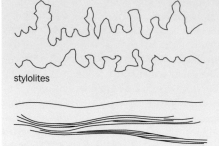

stylolites

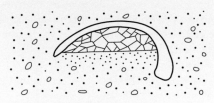

dissolution seams

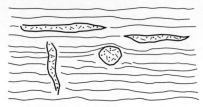

geopetal structure beneath shell

fenestrae in micritic limestone (laminoid, birdseye, and tubular types)

3.26 **Small-scale compaction and cavity features in limestones.**

- Fabric re-arrangement to bed-parallel alignment and more close-spaced packing.
- Sharpening of bedding contacts – especially limestone–mudrock and sandstone–mudrock contacts, and where also associated with pressure dissolution effects.
- Dewatering, fluid migration and diagenetic effects.

Still greater overburden and/or tectonic pressure can lead to pressure dissolution within the sediment. The effects of this are most clearly seen in the field in carbonate and interbedded carbonate/siliciclastic sediments. These include:

- Stylolites – highly irregular dissolution surfaces with insoluble residues (mainly clays) concentrated along them.
- Dissolution seams – sub-parallel to wavy, generally smooth surfaces with concentrations of insoluble residues – leads to bedding appearing wavy to nodular.

A variety of structures result from dissolution leading to cavity formation. This generally occurs either during early burial diagenesis or following uplift and exposure to meteoric waters. Cavity structures (most common in carbonates) are typically infilled or partially infilled with sediment and mineral growth after formation. They include:

- *Geopetal structures:* small cavities (e.g. beneath upturned shells) infilled with sediment in the lower part and cement (e.g. sparry calcite) in the upper part.
- *Fenestrae:* small cavity structures in micritic limestone/dolomite – birdseyes (equant shape), laminoid (lamina cracks parallel to bedding), and tubular (elongate tubes, perpendicular to bedding).
- *Stromatactis:* small cavity with smooth floor of sediment and irregular roof over crystalline (calcite) fill.
- *Sheet crack*s and *neptunian dykes:* larger more continuous (up to many metres), cavities either parallel or transverse to bedding, typically filled with sediment.
- *Karstic cavities:* a whole range of size and shape of cavities formed by meteoric waters in uplifted limestone; potholes, underground streams, caves and caverns, often only partially filled with sediment and speleothems (flowstone, stalactites and stalagmites) (*Plates* 3.77, 7.30).

Observe and measure:
1. The degree of compaction from flattened burrows or laminae thickness variation.
2. The sharpening of lithological contacts that were originally more gradational.
3. The amount of sediment lost through carbonate dissolution – at least equivalent to the cumulative maximum separation of stylolite peaks and troughs.
4. Cavity structures as paleoenvironmental and way-up indicators. They can also be used to determine bedding in limestones where this is obscure.

Chemical weathering

The effects of chemical weathering on exposed rocks in some cases produce pseudostructures that mimic other primary and secondary features:

- *Liesegang rings* are a form of chemical colour banding that may appear like fine lamination, except where they are seen to clearly cross-cut true bedding or lamination. They commonly develop away from joints and other zones of fluid migration through porous sediments. They can be wholly concentric in outline (*Plate* 3.135).
- *Dendrites* are a chemical precipitate of MnO_2 on sediment surfaces exposed to subaerial weathering that takes on a fern-like pseudo-plant fossil aspect. These are most common on bedding planes, joint planes, and prominent faces of exposed outcrops (*Plate* 3.136).
- *Colour mottling* is typically irregular, but in some cases may serve to enhance features such as burrows, bioturbation, lamination, joints, or fissures (*Plates* 3.132, 6.19, 7.33).
- *Iron staining* is one of the most widespread forms of chemical weathering. Iron is readily dissolved by groundwater percolation over iron-rich minerals and then, typically, precipitated as iron oxides and hydroxides.

Chemogenic structures

3.124 Hardground (dashed line) developed in subaqueous environment between successive limestone units. Thin horizon shows iron staining, minor erosion of the underlying beds, burrowing, and possible cavity collapse feature as indicated near lens cap. Lens cap 6cm. *Cretaceous, near Benidorm, SE Spain.*

3.125 Carbonate concretions occurring as bed parallel to sub-parallel lenses and discontinuous layers, within shallow-water sandstone succession. Width of view 10m. *Jurassic, Bridport, S England.*

3.126 Giant carbonate concretion (also known as a dogger), now eroded and fallen from shallow-marine sandstone succession. Width of view 1.8m. Photo by Claire Ashford. *Jurassic, Osmington Mills, S England.*

3.127 Carbonate concretions scattered through deep-water massive (turbidite) sandstone unit; possible orientation along dewatering pathways at oblique angle to bedding.
Miocene, Urbania, central Italy.

3.128 Large carbonate concretion within parallel-laminated mudstone succession. Note that subsequent compaction has caused lamination to bend into and around the earlier cemented concretion. Hammer 25cm.
Cretaceous, central California, USA.

3.129 Small sideritic concretions (brown coloured) in bioturbated volcaniclastic hemipelagic slope succession. Coin 2.5cm.
Miocene, Miura Basin, south central Japan.

3.130 Concretion broken open to reveal ammonite-rich core, hemipelagic slope mudstones. Coin 2.5cm.
Jurassic, Los Molles, west central Chile.

3.131 Bedding-plane view of large, irregular carbonate concretions formed around extensive *Thalassinoides* burrow network in muddy limestone. Width of view 1m.
Jurassic, Osmington Mills, S England.

3.132 Buff/greenish-coloured chemical reduction zones within cross-laminated red bed succession. Green colour from reduced iron (FeII); red colour from oxidized iron (FeIII).
Lens cap 6cm.
Permo–Triassic, Nottingham, central England.

3.133 Wavy/lenticular bedding in micrites, due to pressure-induced dissolution during burial. Width of view 20cm. *Paleogene, Kalavasos, S Cyprus.*

3.134 Fractured, nodular micrite with complex stylolite development due to pressure-induced dissolution during burial and subsequent tectonic stress. Coin 2.5cm. *Cretaceous, near Malaga, SE Spain.*

3.135 Liesegang rings – a chemical segregation of iron oxides and other minerals during weathering. Very common in sandstones, also found in other sediments such as these volcaniclastites, and can be confused with lamination when parallel or subparallel to bedding. Width of view 60cm. *Modern weathering feature, Miocene sediments, Cabo de Gata, SE Spain.*

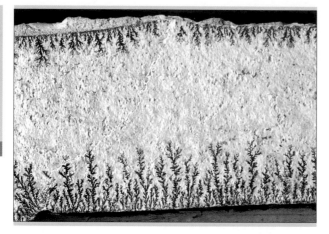

3.136 Dendrites – common pseudofossils occurring on bedding planes of all sediment facies. The intricate plant-like pattern is, in fact, formed chemically during weathering by the systematic precipitation of minute MnO_2 crystals. Bedding plane, width of view 8cm. *Modern weathering feature, Miocene volcanic ash, near Ankara, Turkey.*

3.137 Cone-in-cone structure. This is quite common as a recrystallization form of $CaCO_3$ in carbonate rocks undergoing deep burial diagenesis. Width of view 8cm. *Carboniferous (Mississippian), Derbyshire, UK.*

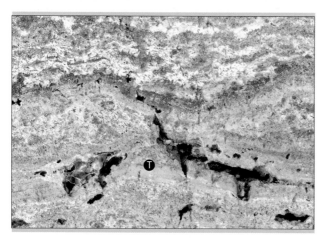

3.138 Irregularly laminated limestone (white) and gypsum (grey) showing laminar, birds-eye, and irregular fenestrae (cavity structures), some partially filled with gypsum and/or sparry calcite. The mounded form (T) in lower part of view is a small-scale tepee structure, typical of lagoonal carbonate and evaporite deposits. Width of view 5cm. *Jurassic, Worbarrow Bay, UK*

Sediment texture and fabric

SEDIMENT TEXTURE refers to the particle grain size and its distribution (e.g. sorting), the grain morphology and surface features, and the porosity–permeability characteristics of the sediment. All these are very important properties that can be used for interpreting the processes and environments of deposition, and for determining fluid flow properties through the rock. Detailed study of sediment texture is generally carried out in the laboratory, but preliminary observations should be made in the field. Use of a hand lens, ruler and grain-size comparator charts is essential (*Figs. 3.27, 3.28*).

Grain size

The mean grain size of siliciclastic sediments is used for classification and nomenclature according to the scheme of Udden and Wentworth, (*Fig. 3.27*). In general terms, grain size reflects:

- The hydraulic energy of the environment – coarser sediments are deposited by swifter currents than finer sediments; mudrocks accumulate in low energy environments.
- The distance from source – mean size decreases downstream in alluvial–fluvial systems, downslope in deep-water, and downwind for explosive volcaniclastic material – but, remember, low energy environments also occur adjacent to high energy conduits, e.g. levee vs. channel.
- The sedimentary material available at source – erosion of friable sandstone will yield only sand and silt grades, oolitic and peloidal limestones are mostly of uniform size, and the size of organic hard parts affects grain size in bioclastic limestones.

Terms for describing grain size of crystalline sedimentary rocks follow those for sands (0.063–2mm), from very finely to very coarsely crystalline. Silt sizes are referred to as

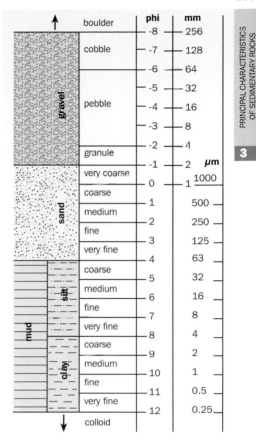

3.27 **Grain-size scale and terms for sedimentary rocks. Note that the arithmetic scale in phi units is used for all serious work in manipulating grain-size data, since it has the advantage of making mathematical calculations easier.**
phi = $-\log_2 d$, **where d = grain size in mm.**
Compiled from original sources.

microcrystalline and clay sizes as cryptocrystalline. Many carbonates, welded pyroclastics, and chemogenic sediments are dominantly crystalline – the crystal size commonly reflecting diagenetic rather than depositional processes. Where original crystal size has been preserved in evaporites, then the larger crystals commonly reflect quieter conditions allowing longer time for their growth.

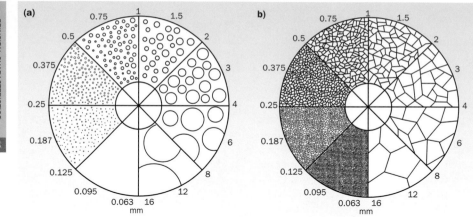

3.28 Comparator chart for estimating the grain size of (a) sands, and (b) crystalline sediments. Place a small piece of the rock or some grains scraped off the rock in the central circle and use a hand lens to compare size with the chart. *Modified after Tucker 1996*

Sorting

Sorting is a measure of the standard deviation from or spread of grain-size distribution about the mean size. It can be estimated in the field from comparator charts (*Fig. 3.29*). In general terms, sorting reflects:

- The degree of agitation and reworking – long periods of transport and/or agitation lead to better sorting, whereas rapid deposition results in poorer sorting.
- The original size and sorting of the source material.
- The grain size of the sediment itself – gravels and muds are generally more poorly sorted than sands.

Grading

Regular variation of grain size through a bed (or part of a bed) is known as grading. The various types of grading are described as sedimentary structures (page 56, *Fig. 3.12*)

Porosity and permeability

Porosity is a measure of the pore space within a sediment.

$$\text{Absolute porosity} = \frac{(\text{bulk volume} - \text{solid volume})}{\text{bulk volume}} \times 100\%$$

$$\text{Effective porosity} = \frac{\text{interconnected pore volume}}{\text{bulk volume}} \times 100\%$$

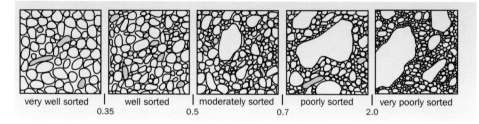

3.29 Comparator chart for estimating sorting in sediments. Numbers are sorting values – i.e. standard deviations expressed in phi units. *Modified after Compton, 1962, Manual of Field Geology, Wiley.*

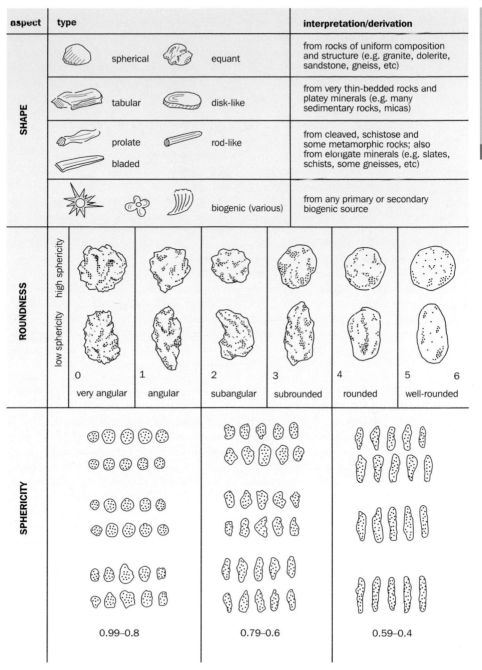

3.30 Comparator charts for estimating the three aspects of grain morphology – shape, roundness, and sphericity.

Permeability is a measure of the ability of a rock to transmit fluids. Both properties, though very important for hydrogeology and petroleum geology, are difficult to measure in the field without the aid of more sophisticated equipment (porosimeters and permeameters). Some estimate can be made by observing the ability of hand specimens to absorb water.

Primary porosity develops as a sediment is deposited and includes interparticle, intraparticle and framework porosity. Primary porosity is closely related to other textural properties. It is highest in fine-grained sediments (around 80% in clastic and carbonate muds), although these show low permeabilities due to capillary restriction at pore throats. Gravels show both high porosity (around 60–70%) and high permeability. Sandstones show good porosities (40–50%) where well sorted, loosely packed, and with rounded, equant grains, and decreasing values with lower textural maturity.

Primary porosity decreases systematically with increasing burial depth as a result of compaction and cementation. Typical values at 4km depth are 5% for immature lithic sandstones to 15% for mature quartzose sandstones. However, secondary porosity can develop as a result of dissolution and dolomitization (the latter in carbonates). Porosity values necessary for good hydrocarbon reservoirs and groundwater aquifers are around 20–35%.

Grain morphology
(*Fig. 3.30, 3.31*)
Three aspects of grain morphology should be considered in the field:

- *Shape (or form)*: determined by the ratios of short, long and intermediate axes.
- *Sphericity:* a measure of how closely the grain approaches an equant or spherical shape.
- *Roundness:* determined by the degree of curvature of grain protrusions.

Comparator charts give a useful field estimate of these parameters. More detailed measurements can be made to give a specific calculation of sphericity and roundness. In general, the shape and sphericity of pebbles reflects their composition and any planes of weakness, whereas their roundness reflects the degree of reworking and/or transport.

Sediment fabric
(*Fig. 3.31*)
Two aspects of sediment fabric (i.e. grain arrangement) should be considered in the field: orientation of grains, and grain packing. Both may result from depositional processes and/or through subsequent burial and tectonic forces.

- Clast and grain alignment of tabular or platy particles sub-parallel to bedding is very common in most sediments.
- Preferred orientation of elongate particles is common in many sediments as a result of long axis alignment either parallel (more common) or perpendicular to flow. Re-orientation can then result from tectonic deformation. Determining the orientation in finer grained sediments requires laboratory measurement.
- Grain imbrication of tabular and disc-shaped particles results from the overlap stacking of grains/clasts such that the direction of dip faces upstream.
- Growth orientation of biogenic material can be preserved where there has been little subsequent erosion and reworking. Transported bioclasts, however, may yield chaotic or preferred orientations.
- Clast/grain-supported fabric is a packing arrangement in which the grains or clasts are in contact with each other.
- Matrix-supported fabric is where the grains or clasts 'float' in an excess of finer-grained matrix (not cement).
- More detailed observations of packing fabrics for sandstones and mudrocks requires laboratory analysis.

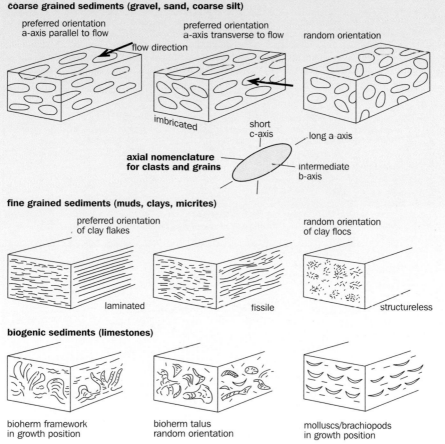

coarse grained sediments (gravel, sand, coarse silt)

preferred orientation
a-axis parallel to flow

preferred orientation
a-axis transverse to flow

random orientation

flow direction

imbricated

short
c-axis

long a axis

axial nomenclature
for clasts and grains

intermediate
b-axis

fine grained sediments (muds, clays, micrites)

preferred orientation
of clay flakes

random orientation
of clay flocs

laminated

fissile

structureless

biogenic sediments (limestones)

bioherm framework
in growth position

bioherm talus
random orientation

molluscs/brachiopods
in growth position

3.31 **Sediment fabrics. (Fabric of fine-grained sediments is only visible under the microscope.)**

Grain surface texture

The grain surfaces of sand and gravel grade material can be examined, at least initially, in the field. Typical surface textures of pebble-size clasts include:

- Dull, polished, smooth – wind-blasted desert clasts.
- Rough, conchoidal fractures, striations – glacial abrasion.
- Smooth, crescentic impact marks – beaches and streams.

High magnification hand lenses can sometimes reveal surface textures of sand grains that are more readily apparent with scanning electron microscopy:

- Dull, frosted – desert sands.
- Minute conchoidal fractures and striations – glacial sands.
- Minute V-shaped percussion marks – beach sands.

Note that diagenetic overprint of surface texture is common.

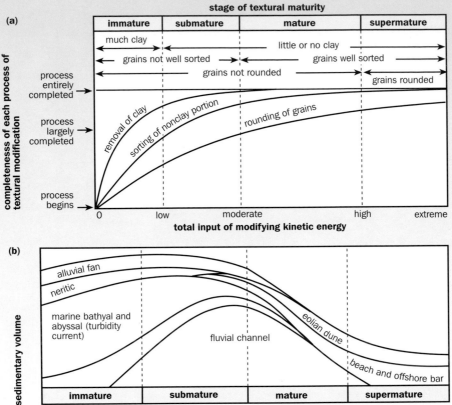

3.32 **The concept of textural maturity in sandstones (a) and particular environments with sands of a characteristic maturity (b).** *Modified after Folk, 1974.*

Observe and measure

1. Estimate mean grain size and sorting, using charts and/or tape measure as necessary; note maximum grain or clast size and whether there is a single mode or bimodal/polymodal distribution.
2. Note any graded beds, and systematic vertical and lateral changes in grain size or sorting.
3. Try to make some estimate of porosity–permeability.
4. Make appropriate estimates of grain morphology attributes – shape and roundness, in particular; look also for distinctive grain surface textures.

5. Note any preferred orientation of elongate grains/clasts and take compass bearing; note imbrication and direction of dip of clasts; determine whether biogenic material is in growth or non-growth position.
6. Determine whether sediment is grain-supported or matrix-supported.
7. Try to estimate textural maturity of the sediment – using *Fig. 332*.

Sediment composition

SEDIMENT COMPOSITION refers to the nature of the components that make up the sedimentary rock. Broadly speaking, this comprises the framework or coarser particulate material, the matrix, or finer particulate material, and cement – including individual authigenic minerals as well as binding material formed during diagenesis.

Practically any mineral or rock fragment, fossil animal or plant remains can occur in sediments (Tables 3.1 and 3.2). The main classes of sedimentary rocks are based largely on sediment composition: terrigenous, biogenic, chemogenic and volcaniclastic. Sub-

Table 3.1 Principal components of terrigenous sediments and sedimentary rocks (conglomerates, sandstones, and mudrocks)

Minerals

Quartz	Includes monocystalline and polycrystalline quartz.
Feldspars	Include K-feldspars and plagioclase feldspars.
Clay minerals	Include kaolinites, illites, smectites, chlorites and sepiolite.
Glauconitic material	Commonly mineral admixture including glauconite.
Mica minerals	Include muscovite (sericite) and biotite.
Carbonate minerals	Include calcite, dolomite, siderite and ankerite.
Evaporite minerals	Include gypsum, anhydrite and others.
Heavy minerals (non-opaque)	Include large number of typical rock-forming minerals with specific gravity >2.9.
Heavy minerals (opaque)	Include hematite, limonite, goethite, magnetite, ilmenite, leucoxene, pyrite, marcasite.
Zeolites	Common alteration products of volcanic glass.

Biogenic and organic material

Calcareous skeletal matter	
Siliceous skeletal matter	
Phosphatic material	Including apatite.
Organic matter	Includes palynomorphs, coaly material and kerogens.

Rock fragments

Igneous rocks	Include all plutonic and volcanic rock types.
Metamorphic rocks	Include all metamorphic types.
Sedimentary rocks	Include all types of sediment.

Cementing material

Silicates	Mainly quartz, also chalcedony, opal, feldspars, zeolites.
Carbonates	Mainly calcite, also aragonite, dolomite, siderite.
Iron oxides	Hematite, limonite, goethite
Sulphates	Anhydrite, gypsum, barite, celestite

divisions within these groups, apart from the grain-size divisions of terrigenous clastics, are also commonly made on compositional attributes. The topic is therefore discussed more thoroughly in the individual lithology chapters (Chapters 4–14). The common properties of sedimentary grains under a hand lens are given in Table 3.3 and illustrated in *Plates 3.124–3.139*.

Inorganic components

Of the inorganic components that are dominant in non-biogenic sediments, the chemically more stable components are less readily broken down during weathering and hence more commonly preserved in the rock record.

The concept of compositional maturity as a reflection of component stability during the weathering, transport, and post-depositional processes is illustrated in *Fig. 3.33*. The other principal factors which determine the relative abundance of different components in sedimentary rocks are the original abundance of different rock and mineral types in the source area, and the nature of the depositional area. Not only will diagenetic processes remove certain unstable minerals from sediments, but they also lead to the addition of cementing minerals in the available pore spaces – quartz and calcite are the dominant cements, but others also occur (Table 3.1).

Observe and measure

1. The framework grains or clasts – noting principal, minor and accessory types and estimating percentages.
2. The matrix composition – although this can be more difficult to ascertain even using a hand lens.
3. The nature of the cementing material – and any observations on pore filling or replacement types.

Organic and biogenic components

For convenience, the terms organic and biogenic are used to refer to different parts that an organism has contributed to the sediment.

Table 3.2 Principal components of biogenic sediments and sedimentary rocks

Calcareous skeletal matter

Calcareous algae and microbial precipitates – include red and green algae, stromatolites and coccoliths

Foraminifers – include planktonic and benthic forms

Coelenterates – mostly corals

Sponges – include stromatoporoids

Bryozoans

Brachiopods

Molluscs – include bivalves, gastropods, pteropods, cephalopods

Echinoderms

Arthropods – include ostracods and trilobites

Siliceous skeletal matter

Diatoms

Radiolarians

Sponge spicules

Silicoflagellates

Non-skeletal carbonate grains

Ooids and pisoids

Peloids

Aggregates

Intraclasts – include variety of pre-existing biogenic sediments

Lime mud or micrite

Phosphatic material

Fragments of bones, teeth and scales

Carbonaceous material

Woody debris

Pollen and spores

Kerogen

Cementing material

Calcite group – mainly calcite, also siderite, magnesite, rhodochrosite

Aragonite group – mainly aragonite

Dolomite group – mainly dolomite, also ankerite

Organic refers to all that material derived from soft organic tissues and higher plant material (leaves, wood, etc). By the time it has become part of a sedimentary rock, it has variously altered by thermal and bacterial degradation, yielding yellow-brown amorphous and particulate kerogens, liquid hydrocarbons (oil), and brown-black carbonized plant debris and coal. In all of this material, organic carbon remains the principal constituent. Oil is generally detected in sediments by its strong petroliferous odour and yellow-brown colour. Most of the light hydrocarbon fraction (oil and gas) will have leaked away long before the sedimentary rock reaches the surface. Very heavy oils and residual oil can remain behind as a brown to black tar-like impregnation, sometimes giving soft pliable sands (tar sands).

Biogenic refers to the hard skeletal remains of organisms that find their way into sedimentary rocks of all types. They may be fragmented and altered by diagenetic processes, or perfectly preserved as recognizable body fossils. Microfossils are often the dominant component of many fine-grained carbonate sediments, but they are not easy to observe. Coccoliths, for example, make up chalks and micrites the world over, but are so tiny that

FIELD TECHNIQUES
To determine sediment composition in the field, it is essential to examine the rocks carefully with a hand lens. Hammer off a fresh specimen or place your nose close to the outcrop, and use a 10x or 15x lens. Familiarize yourself with some of the principal components viewed this way by reference to the following pages (*Plates 3.139–3.134*). Try to estimate approximate percentages of key components, distinguishing between grains and cement.

they can only be observed under high power microscopes. Some radiolarians and diatoms are silt-sized microfossils whereas other species, together with most foraminifers, are sand-sized. These can be recognized by careful use of a hand lens. They are sometimes easier to identify on weathered rather than fresh rock surfaces. Macrofossils are more easily found and described in the field as part of the sediment composition.

3.33 Concept of compositional maturity in sandstones showing the components most characteristic of different maturity fields.

component	immature	submature	mature	supermature
quartz			poly-crystalline	mono-crystalline
feldspar	(Ca) plagioclase (Na)	orthoclase microcline		
micas	biotite	chlorite	muscovite	
heavy minerals	olivine amphiboles	pyroxenes glauconite	garnets	zircon rutile tourmaline
opaques				
rock fragments	schist phyllite slate			
(volcanics)	basic ——————— intermediate ——————— acid			
	evaporites	←————ironstones————→		
	←————————limestone / dolomite————————→			
biogenics	←— — — — — — — — — —→			
plant fragments	←— — — — — — — — — —→			
	increasing chemical attack			
	during weathering, transport, early – late diagenesis ————————————————→			

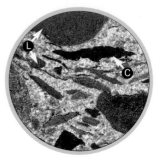

Hand lens photographs of sedimentary grains

◀ 3.139 Polymict conglomerate with quartz (Q), volcanic (V), and schist (S) clasts. Note brown iron-staining on one clast.

▶ 3.140 Sandy oligomict conglomerate with carbonaceous (C) and lithic (L, siltstone) clasts in a feldspar-rich matrix.

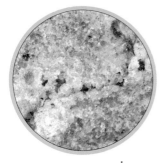

◀ 3.141 Pebbly feldspathic muddy sandstone (greywacke), with lithic (L), quartz (Q), and feldspar (F) clasts/grains, and greenish chloritic mud matrix.

▶ 3.142 Carbonaceous muddy sandstone (greywacke) with large woody fragments (black) in fine-grained matrix.

◀ 3.143 Quartz sandstone (quartzarenite) with well sorted, well-cemented, grey/white quartz grains, minor opaques (black) and iron-staining.

▶ 3.144 Red-stained feldspathic sandstone (arenite), with well rounded, hematite-coated, quartz, white feldspar, and rare, dark, lithic grains.

◀ 3.145 Feldspathic sandstone (arkose) with grey translucent quartz, white feldspar, rare black opaques, and minor iron-stained matrix.

▶ 3.146 Glauconitic sandstone with quartz (grey, translucent white), feldspar (white), and green-black glauconite.

◀ 3.147 Biosparite (skeletal grainstone) with circular crinoid ossicles replaced by coarse sparry calcite.

▶ 3.148 Bio-microsparite (skeletal packstone–grainstone) with aligned broken bivalves in fine microspar cement. Green chloritic clay.

◀ 3.149 Oosparite (ooidal grainstone) packed with spherical ooids, some now weathered partly hollow, and sparite cement.

▶ 3.150 Oosparite (ooidal grainstone) with spherical ooids, rare bioclasts, and much interstitial sparite cement.

◀ 3.151 Calcareous ironstone with white, degrading carbonate grains (some ooidal) in matrix of iron-rich silt (brown) and mud (green).

▶ 3.152 Chamositic oolite with iron-stained and degraded carbonate/chamositic ooids and fine iron-rich matrix. Note stylolite development (S).

◀ 3.153 Anhydrite, as white sugar-textured matrix, with isolated dark, euhedral–subhedral gypsum crystals.

▶ 3.154 Microbial laminated limestone and evaporites, with original gypsum now replaced by quartz (Q). Dark bands are organic-rich.

Fossils

FOSSILS are one of the most important components of sedimentary rocks and deserve special attention. They are the key to biostratigraphy, for the dating and correlation of rocks, although this generally requires specialist knowledge, careful collecting and laboratory work. Equally, they provide vital clues for the environmental interpretation of sediments, providing information on: the marine, freshwater or terrestrial nature of the depositional environment; the water depth, agitation, turbulence and flow energy; sedimentation rate, resedimentation and winnowing; and general paleoclimate data.

Fossil assemblages closely reflect the depositional environment. Many different types of assemblage are easily recognized: an *in-situ* bioherm assemblage with corals, algae and bivalves; a jumbled reef talus assemblage with similar but fragmented forms; an oyster bank assemblage dominated by one species of thick-walled bivalve; a thin-shelled echinoid–bivalve–sponge assemblage characteristic of deeper quiet seas; a terrestrial plant–freshwater fish assemblage; and so on. Always be aware that most of the fossil record is not preserved, so that only a partial picture emerges. Use trace fossils and the presence of encrusting or boring organisms to supplement this picture (page 86).

Species diversity also depends on environmental factors, such as water depth, salinity, oxygen levels, agitation, and substrate. Where conditions are optimal then species diversity is at a maximum, with many different types of fossils and trace fossils. Where conditions are stressed then species diversity is lower, although one particular species may become very abundant – this is especially true of salinity extremes. With increasing water depths, pelagic and nektonic fossils dominate (e.g. cephalopods, graptolites, fish, many microfossils, and ostracods). The same occurs with decreasing oxygen content of bottom waters, so that benthic and burrowing organisms gradually decrease and disappear altogether. Euryhaline forms are those species able to tolerate extremes of salinity – e.g. certain ostracods, gastropods, bivalves, and algae. Stenohaline forms require normal marine salinities – e.g. corals, bryozoans, trilobites, and many other species of a whole range of fossil groups.

Fossil distribution is an important variable. Evenly dispersed fossils are typical of quiet conditions of pelagic accumulation. More commonly, concentrations occur either as *in-situ* accumulations (e.g. bioherms, shell banks) in which fossils remain in their growth position, or as current accumulations. Many different types of current accumulation exist, formed by fluvial or tidal reworking, shelf currents and storm events, deep-water bottom current winnowing (mainly of microfossils), resedimentation in turbidity currents, and as bioherm talus slopes.

Fossil preservation varies according to both depositional and diagenetic conditions. The higher the degree of agitation and transport during current accumulation, the greater the degree of fragmentation and eventual rounding that occurs. Fossils may also become mineral-coated (especially iron-stained) during this process. Many fossils are replaced by minerals other than the original during diagenesis – calcite, dolomite, silica, pyrite, and hematite are common occurrences. Many more dissolve and disappear completely, in some cases giving rise to concretion formation and pervasive cementation.

Observe and measure

1. *Fossil assemblages:* the number, types, and relationships of fossils present. Note also the variation between different beds or parts of the succession.
2. *Species diversity:* the number and type of species present.
3. *Distribution of fossils:* the location and occurrence of fossils within the sedimentary section, and whether they represent *in-situ* or current accumulations.

4. *Preservation of fossils:* the degree of fragmentation, rounding, mineral coating, diagenetic alteration and replacement. With careful hunting and use of hand lens, ghosts of fossils may be revealed. Early-formed concretions sometimes show better preservation of fossils than the surrounding sediment.

Sediment colour

COLOUR is one of the most conspicuous properties of sedimentary rocks in the field, and commonly figures in both our informal field descriptions and more formal stratigraphic nomenclature. It is an important property to observe, taking care to distinguish between the colour of weathered surfaces, much affected by surface processes and lichen, and fresh surfaces, that are a better reflection of depositional or diagenetic environments. For a discussion of chemical weathering and its effect on sediment colour, see page 103. For careful work, colours should be recorded by comparison with the *Munsell Soil Colour Chart* or its derivative, the *Rock Colour Chart*.

The principal controls on colour are the organic carbon content, the oxidation state of iron and the colour of the principal rock-forming minerals (*Fig. 3.34*). The progression of colours from light grey through dark grey to black correlates with increasing carbon content. Dark grey to black colours can also result from trace amounts of finely dispersed iron sulfides (pyrite and monosulfides), hybrid Fe^{2+}/Fe^{3+} minerals (magnetite, ilmenite), and Mn^{4+} (manganese dioxide). Colours from red through purple to green and grey correlate with decreasing Fe^{3+}/Fe^{2+} ratio. Lighter colours (white to very pale grey) also correlate with very high carbonate or silica content, in the absence of organic carbon or other impurities. Green, yellow, orange, and mauve colours derive from variable iron impurities, whereas pinkish colours can result from both iron and Mn^{2+}.

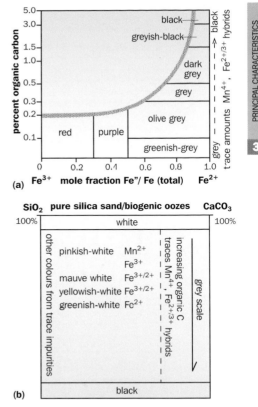

(a)

(b)

3.34 **Chemical controls on colour in sediments: (a) based on the suggested relationship of mudrock colour to carbon content and the oxidation state of iron** (*modified after Potter et al 1980*) **but note that grey-black colouration can also result from trace amounts of Mn^{4+} and $Fe^{2+/3+}$ hybrids; (b) deviation from the white colour of pure silica sands (quartz sandstone) and biogenic ooze (limestone), due to trace contents of the elements shown.**

Colour is therefore closely associated with redox conditions both in the depositional environment and during diagenesis. Quieter, more reducing conditions will favour preservation of organic carbon and hence darker coloured sediments (black shales) typically in marine or lacustrine settings; open-water marine and terrestrial (alluvial–fluvial) oxidizing conditions will lead to oxidation of Fe^{2+} to Fe^{3+} and hence 'red bed' sediments.

Table 3.3   **Common properties of sedimentary components under a hand lens for field identification of sediments and sedimentary rocks**

Minerals (grains and cements)

Quartz	Ubiquitous in terrigenous sediments, 65% of average sandstone (up to 99%), 30% of average mudrock and conglomerate, 5% of average carbonate rock, present in some volcaniclastics. Vitreous lustre (glassy), pale grey to sub-transluscent, but may be slightly coloured (pink, green, yellow) with impurities; lacks alteration and cleavage, rounded to angular grains, sub-equant (more elongate in silt size); H7 – not scratched with penknife, scratched by steel file.
Feldspars	Common in terrigenous sediments, 10–15% of average sandstone and conglomerate, 5% of average mudrock, <1% of average carbonate rock, common in many volcaniclastics. Vitreous/sub-vitreous lustre, white or pinkish (also greenish, brownish), may be more dirty and ragged due to alteration; generally more rounded than quartz, but good cleavage typically gives prismatic or tabular grain shapes; H6 – just scratched with penknife and by quartz.
Clays	Dominant minerals of mudrocks, very variable in sandstones, conglomerates and carbonate rocks, very common alteration product in volcaniclastics. Individual grains cannot be distinguished with the naked eye; forms matrix in the coarser-grained rocks, the darker-coloured layers in laminated sediments and the bulk of fine-grained sediments; typically grey, dark grey and greenish colours (also reddish, purplish, black); dark grey and black colours commonly indicate organic-rich sediments, greenish colours may indicate chloritic material as an alteration product.
Glauconite	A clay mineral of the illite group that commonly occurs as part of a complex mineral admixture in greenish sand-sized grains (strictly glauconitic grains); generally smooth spherical, ovoid to lobate in form, also irregular and other shapes; green, dark green, greenish-black when fresh, but oxidizes readily to a dark rusty brown. H2 – scratched easily with fingernail.
Micas	Common accessory minerals in terrigenous sediments, but very visible due to their sparkly appearance in sunlight; muscovite and sericite (its alteration product) are the most resistant and hence common, whereas biotite only occurs in immature sediments. Pearly lustre, flat faces typically sparkle brightly in sunlight, darker as edge-on grains; colourless, white, brown or greenish flakes; flat, platy, ragged shape, may be conspicuously larger than associated grains; H2.5 – just scratched with fingernail.
Carbonate minerals	Calcite dominant in limestones, dolomite dominant in dolomites, aragonite is the unstable form of calcite common in modern calcareous sediments, siderite may form as brownish-coloured nodules in mudrocks.
Calcite	Vitreous lustre, colourless but typically cloudy with impurities giving whitish, greyish colours (also pink, yellow, green); good crystal shape with three cleavages at 60° giving part rhombs, also fibrous varieties; micrite is a micro-crystalline calcite in which grains are indistinguishable even under a hand lens; sparite is a coarser crystalline form that typically sparkles brightly in sunlight. H3 – scratched with copper coin; vigorous effervescence with dilute acid.
Dolomite	Vitreous lustre, often pinkish, brownish or orange due to Fe impurities; strong crystal form giving rhomb shape; will only effervesce when scratched with knife and powdered.

Table 3.3   ...continued

Evaporite minerals	Evaporite minerals – dominant in evaporite deposits, but also occur in carbonate rocks, especially dolomite, and in mudrocks; highly soluble so that grains and crystals may easily be replaced by pseudomorphs.
Gypsum	Vitreous lustre, colourless, transluscent, but also cloudy white to grey with impurities; typically crystalline, fibrous or swallow-tail crystal shape. H2 most common evaporite – easily scratched with fingernail.
Anhydrite	Vitreous lustre, colourless, but more commonly white to grey with impurities; typically fine crystal size giving granular, sugary appearance. H4 relatively rare at the surface – scratched with glass, and easily scratched with knife.
Halite	Vitreous lustre, colourless to white (other colours due to inclusions); coarsely crystalline, often as well-defined cubes or part cubes, but may appear massive. H2.5 – just scratched with fingernail, distinctly salty taste. Sylvite co-occurs with halite but tastes bitter.
Barite	Has a high specific gravity. Barite, celestite, strontianite, polyhalite and other exotic evaporites are rare in occurrence.
Heavy minerals (non-opaque, SG>2.9)	At least 100 different species have been noted in sedimentary rocks, may stand out because of their high relief but mostly difficult to distinguish under a hand lens. More common species listed below in approximate order of their chemical stability (most stable first).
Zircon	Euhedral–subhedral small prismatic to anhedral equant, colourless.
Rutile	Euhedral–subhedral small prismatic to rounded, red-brown.
Tourmaline	Euhedral–subhedral elongate prismatic to subrounded, brown-bluish.
Sphene	Subhedral diamond-shaped to anhedral irregular, colourless – yellow.
Spinel	Subhedral well-rounded, colourless, green, brown varieties.
Garnet	Anhedral equant, colourless, pink, red, yellow varieties.
Cassiterite	Euhedral–subhedral small prismatic – pyramidal, brown – red.
Kyanite	Euhedral–subhedral small rectangular, colourless.
Andalusite	Subhedral prismatic to anhedral irregular, colourless – pinkish.
Sillimanite	Subhedral prismatic to anhedral irregular, colourless.
Apatite	Euhedral prismatic, colourless.
Staurolite	Anhedral angular, colourless – straw yellow.
Epidote	Subhedral prismatic to anhedral elongate, yellow-green.
Amphiboles	Varied group, typically subhedral prismatic, elongate or ragged (also fibrous varieties), colourless to green/brown.
Pyroxenes	Varied group, typically subhedral prismatic to anhedral small stubby crystals, colourless – greenish.
Olivine	Anhedral equant subrounded, colourless to green.

Table 3.3   ...continued

Heavy minerals (opaque, SG>2.9)	A number of heavy metal species are also common accessory minerals in many sedimentary rocks; they typically appear as small, black, irregular to subrounded grains with a metallic to sub-metallic lustre. In bright sunlight certain distinctive colours may be visible under a hand lens.
Magnetite	Black, blue-black.
Ilmenite	Blue-black to purple-brown, weathers to white leucoxene.
Hematite	Grey-black to reddish-brown, also as thin coating on many other grains.
Limonite	Dark brown to yellowish-brown, submetallic–earthy.
Pyrite	Brass-yellow, also as cubes, framboids (spheres), worm and fossil casts.
Marcasite	Brass-yellow, fibrous.
Chalcopyrite	Strong brass-yellow, iridescent.
Arsenopyrite	Silver-white to steel-grey.
Micronodules	Black, bluish-grey, greenish-grey or reddish-brown, mainly spherical, also irregular in shape, sub-vitreous to dull; include a range of iron and manganese oxides and hydroxides (including goethite).
Zeolites	Common minerals formed through the alteration of volcanic and volcani-clastic rocks, but of the 33 types only clinoptilolite, phillipsite and analcime occur in sufficient abundance to be recognized. These display a vitreous lustre, colourless very fine crystals that together appear white, also pale yellow due to inclusions or coatings.

Biogenic and organic materials

Calcareous skeletal matter	Calcareous algae and microbial precipitates – many not resolvable under a hand lens except for coralline (red) algae, microbial ('algal') mats and stromatolites in growth position, and as clasts, with distinctive crenulated lamination.
Foraminifers	Planktonic and benthic microfossils, 50–1000μm (rarely larger – e.g. fusulinids), variety of shapes – especially globular and turreted, chambered.
Corals	Many types of corals in growth position; individuals or bioherms, and as transported fragments.
Sponges, bryozoans	Isolated, fragmented or as parts of bioherms, characterized by numerous pores.
Molluscs	Very common as whole fossils or as fragments of tests, fibrous or granular calcite test structure; include bivalves, gastropods, cephalopods.
Arthropods	Include small (0.15–3mm) bivalved ostracods, and larger whole or fragmented trilobites.
Brachiopods	Common in Paleozoic and Mesozoic shallow-marine deposits.
Echinoderms	Common marine fossils; crinoids common Paleozoic–Mesozoic.

Table 3.3   ...continued

Siliceous skeletal matter	Siliceous skeletal matter – originally opaline silica, fine structure of organism lost on recrystallization to quartz; also completely redistributed as cement.
Diatoms	Planktonic and benthic microfossils, $10–100\mu m$ (also up to 1mm), variety of spherical, discoidal and rod-like shapes.
Radiolarians	Planktonic microfossils, $20–400\mu m$, spherical, conical and helmet-shaped forms.
Sponge spicules`	Rare minute needle and stellate shaped fragments.
Silicoflagellates	Rare delicate latticework skeletons, $20–50\mu m$, commonly broken.
Carbonaceous and phosphatic material	Common accessory components of terrestrial and nearshore shelf sediments, also in deep-water turbiditic settings.
Woody debris	Brown-black lignite, some elongate and with surface striations, also carbonized black debris with irregular shapes.
Pollen, spores	Small $20–80\mu m$ grains, yellowish and brownish colours typical, spheroidal, oblate, rod-like and other shapes common.
Kerogen	Amorphous orange-brown material.
Hydrocarbons	Light oils readily migrate out of most sediments, but may leave behind a yellow-brown stain and distinctive smell; heavy oils and bitumens can remain as a thick, soft, sticky, dark-brown to black substance within the pore spaces.

Rock fragments

Igneous rocks	All types are possible, but the most common are fine-grained acid to intermediate volcanic rocks and glasses, and medium-fine grained intrusives. A complete range of extrusive and fragmented igneous material occurs in volcaniclastic sedimentary rocks (see Chapter 14). Palagonite is a yellow-brown (also greenish) marine alteration product of basaltic glass; irregular to shard-like form, commonly vesicular. Coarser igneous rocks may occur close to source, more distally they are typically broken up into individual mineral grains.
Metamorphic rocks	All types are possible, but the most common are fine to medium-grained slates, schists, psammites, and metamorphosed volcanics. Coarser metamorphic rocks may occur close to source, more distally are typically broken up into individual mineral grains.
Sedimentary rocks	All types are possible, but the most common are the well-cemented and chemically stable types, including chert, acid tuffs, silicified limestones, dolomites, micrites, compacted mudrocks (shales), siltstones, and quartz sandstones. A range of other sedimentary rocks can occur in proximal deposits close to the source region, or where transport and deposition has been very rapid.

CONGLOMERATES

Incised alluvial fans, River San Juan, NW Argentina

Definition and range of types

CONGLOMERATES are all those coarse-grained sedimentary rocks that consist dominantly of gravel-sized (>2mm) clasts. They are also known as rudites. Strictly speaking, conglomerates should contain >50% clasts over 2mm in diameter; anything less than this and they are more correctly termed pebbly sandstones or pebbly mudstones as appropriate. Most melanges, olistostromes, and many debrites fall into this category and are therefore discussed in Chapter **6** (*Mudrocks*).

Poorly sorted or non-sorted sediments that contain a wide range of clast sizes (pebbles, cobbles, boulders) in a muddy matrix are also called diamictites. Some authors reserve this term for use with glacially deposited pebbly mudstones and muddy conglomerates (also known as till or tillites). Conglomerates, in which the clasts are separated by finer-grained sediment, are known as matrix-supported, whereas those in which the clasts are in contact with one another are termed clast-supported. Conglomerates with a dominance of angular rather than rounded clasts are known as angular conglomerates, or breccias.

On the basis of clast origin, intraformational and extraformational conglomerates can be distinguished. Intraformational clasts are those derived from within the basin of deposition, and typically include shales or micritic limestones. Extraformational clasts are those derived from outside the basin, and can include a wide range of types. In many cases, of course, it is not possible to determine whether the clasts are intraformational or extraformational, so neither prefix should be used. Polymict conglomerates are those with many different types of clast, oligomict and monomict conglomerates have, respectively, few and just one type of clast. Where the dominant clast type is limestone or dolomite the rock is known as a carbonate conglomerate or calcirudite (Chapter 7). Likewise, where volcanic clasts are dominant the rock is a volcaniclastic conglomerate (Chapter 14).

In some cases, angular conglomerates or breccias are formed *in situ* by breakage, collapse or solution. Such rocks are termed cataclastic breccias and solution breccias. Where the clast size is extremely large, then the term megabreccia or mega-conglomerate can be used, whatever the origin.

The fundamental genetic types of conglomerates are shown in Table 4.1.

Principal sedimentary characteristics

See photographs and figures in the relevant sections of Chapter 3 as well as at the end of this chapter.

Bedding

- *Thickness*: variable, may be very thick (massive); bedding commonly indistinct or absent. Bed thickness may vary systematically through a succession, often in association with sandstone beds, where they are referred to as thinning-up or thickening-up sequences. Conglomerates are also prone to showing proximal to distal decrease in bed thickness over relatively short distances (e.g. tens to hundreds of metres).
- *Shape*: irregular and lenticular beds common, in some cases with channel-like geometry.
- *Boundaries*: top and bottom boundaries typically irregular or gradational; bottom may be erosive and sharp.

Structures

Large clast size and poor sorting commonly make it difficult to observe primary structures in conglomerates. Many beds may appear structureless (or massive) initially but closer inspection can reveal crude (or subtle) stratification – look carefully for parallel-alignment of elongate clasts. Both parallel and cross-stratification occur, with the latter in some cases only slightly inclined. Normal and reverse grading occur through distinct beds, but an irregular oscillation of grain size is often observed in an unbedded or poorly bedded unit. In some conglomerates the larger clasts have been rafted towards the middle or top of the bed due to buoyancy forces acting during transport (e.g. in debris flows). Soft and semi-consolidated sediment clasts typically show deformation structures. Water-escape features are rare and bioturbation generally absent.

Fabric

Matrix-supported and clast-supported fabrics both occur in conglomerates, with clast-support more typical in fluvial, beach, reef-talus and many volcaniclastic deposits, whereas matrix-support is more common in debrites (subaerial or subaqueous) and glacial diamictites. Thin-bedded clast-supported conglomerates, showing evidence of extensive reworking and abrasion, may be due to rapid marine transgression over a low-lying continental shelf. Such basal conglomerates occur at the base of a transgressive system tract. Other fabric types are noted by the disposition of tabular and blade-shaped clasts: e.g. random, bed-parallel or sub-parallel, and imbricated. In fluvial and shallow-marine current-deposited conglomerates, long axes are generally oriented normal-to-current as the result of a rolling action of pebbles over the bed surface. In glacial diamictites, a parallel-to-flow orientation results from a sliding action, while in debrites and coarse-grained turbidites a parallel-to-current orientation results from very rapid deposition and

Table 4.1 Principal types of conglomerates and breccias

Major types	Subtypes	Nature and origin
Conglomerate and breccia	Extraformational con-glomerate and breccia – includes polymict, oligomict and monomict subtypes.	Breakdown of older rocks of any kind through the process of weathering and erosion; deposi-tion by fluid flows (water, ice) and sediment gravity flows.
	Intraformational conglom-erate and breccia – mainly monomict types.	Penecontemporaneous fragmentation of weakly consolidated sedimentary beds; depo-sition by fluid flows and sediment gravity flows.
Volcaniclastic conglomerate and breccia	Pyroclastic conglomerate and breccia	Explosive volcanic eruptions, either magmatic or phreatic (steam) eruptions; deposited by air-falls or pyroclastic flows.
	Autoclastic breccia	Breakup of viscous, partialy congealed lava owing to continued movement of the lava.
	Hyaloclastic breccia	Shattering of hot, coherent magma into glassy fragments owing to contact with water, snow or water-saturated sediment.
	Epiclastic conglomerate and breccia	Reworking of previously deposited volcanic material.
Cataclastic breccia	Landslide and slump breccia	Breakup of rock owing to tensile stresses and impact during sliding and slumping of rock masses.
	Tectonic breccia: fault, fold, crush breccia	Breakage of brittle rock as a result of crustal movements.
	Collapse breccia	Breakage of brittle rock owing to collapse into an opening created by solution or other processes.
	Meteorite impact breccia	Shattering of rock owing to meterorite impact.
	Solution breccia	Insoluble fragments that remain after solution of more soluble material; eg. chert clasts con-centrated by solution of limestone.

Modified from: Pettijohn FJ, Sedimentary Rocks (3rd ed),1975; and Boggs S, 1992

flow freezing. *In-situ* brecciation processes in limestones (e.g. karst collapse, hardground fragmentation, and calcretization of soil horizons), autobrecciation in volcaniclastic deposits (e.g. autoclastites and hyaloclastites), and cataclastic processes (e.g. various tectonic breccias), all tend to produce random clast fabrics.

Texture

Conglomerates have a dominant mean size >2mm, but include a wide range of size and sorting characteristics. Some of the largest boulders or clasts may be the size of a car or house! Modal size is generally easier to determine than mean size in the field and many conglomerates are, in fact, bimodal or polymodal in their grain size distribution. The maximum clast size is also a good indication of flow strength or velocity. With some depositional processes there is a positive correlation between maximum clast size and bed thickness – this is true of muddy debris flows and stream floods. Most other processes, including normal braided river flow, do not yield this relationship. Depositional porosity and permeability are generally very high, except where there is an abundant muddy matrix. Both decrease markedly with compaction and cementation.

Composition

Conglomerates, like sandstones, can include almost any pre-existing mineral or rock fragment (see **Table 3.1**), the least stable ones only being preserved where deposited close to source. Field observations should include the type, variety and approximate proportions of different rock types present as clasts. These data give very good information on the provenance (source area) and likely transport distance. Different levels or beds may yield different compositions, suggesting change of source area or multiple sources.

4.1 Classification of conglomerates based on clast type and matrix content. Note that within each of the principal types (metamorphic, igneous and sedimentary) it is possible to have monomict, oligomict and polymict varieties. Breccia is the term used where the dominant clast shape is angular.

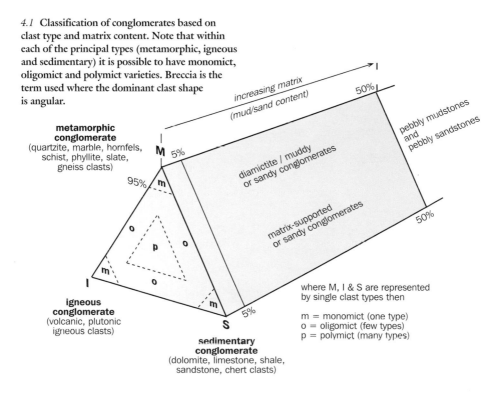

Classification of conglomerates

THERE ARE relatively few classification schemes for conglomerates, apart from those encompassed in the definition of types listed above. Clast size can be used as a descriptor term such as cobble-rich or boulder-rich conglomerate, but a more systematic compositional classification is probably most helpful. Terms for the range of different clast types have already been given; terms for the relative clast stability also exist. Conglomerates made up of framework grains that consist dominantly of ultrastable clasts (i.e. >90% quartzite, chert, and vein quartz) are quartzose conglomerates. Those with abundant metastable or unstable clasts are petromict conglomerates.

The classification scheme proposed by Boggs (1992) is currently the most comprehensive, and this has been modified herein (*Fig. 4.1*).

Occurrence

CONGLOMERATES and breccias are deposited in a range of environments by high-energy processes. They are typical of continental

FIELD TECHNIQUES
Mark out a 0.5m by 0.5m quadrat (or equivalent) on the rock surface. Measure grain size, clast orientation and composition for at least 50 clasts (200 will give better statistics). Plot clast orientation on a rose diagram, after stereonet correction for dip of beds if over 10–15°. Plot clast composition on a bar chart or pie diagram. Both bedding plane and vertical-to-bedding surfaces must be examined for orientation data.

environments, such as those of alluvial fan and fluvial systems, where they may occur as part of a red-bed succession. They also occur in glacial deposits, typically as matrix-supported conglomerates and pebbly mudstones, or on fan-deltas just fringing into a lacustrine or marine setting. Thinner deposits occur in beach and shallow marine settings, where they are associated with shallow water fossils, calcareous encrustations and borings. In deeper water, slope apron, and submarine fan systems, they are common deposits of debris flows and high-concentration turbidity currents, especially in submarine channels.

Field photographs

4.1 Polymict conglomerate (breccia) showing clast alignment and reverse grading; part of deep-water turbidite succession.
Cretaceous, Dana Point, California, USA.

4.2   Monomict limestone conglomerate (breccia) with random clast arrangement in sparry calcite cement; part of proximal rockfall scree deposit. Wine pouch 15cm wide.
Plio–Pleistocene, near Benidorm, SE Spain.

4.3   Sandy matrix-supported polymict conglomerate, poorly sorted with clasts up to 50cm diameter in 1m thick bed; sandy debrite within deep-water turbidite succession. Hammer 45cm.
Eocene, Tabernas Basin, SE Spain.

4.4   Polymict clast-supported conglomerate, very poorly sorted with clasts over 1.2m in diameter, bedding indistinct to absent, crude stratification due to subparallel clast alignment (tabular clasts), indicate bedding dips at approximately 35° from upper right to lower left; part of coarse-grained alluvial fan system.
Neogene, NW Crete, Greece.

4.5 Monomict limestone conglomerate, very poorly sorted with clasts over 1.5m diameter, crude subparallel clast alignment but no clear bedding; part of debris flow/flow–slide deposit on subaerial alluvial fan. Bedding is approximately horizontal, based on basal contact of this unit (out of view). Hammer 30cm.
Plio–Pleistocene, near Benidorm, SE Spain.

4.6 Monomict limestone conglomerate, poorly defined bedding, crude stratification due to subparallel clast alignment, but note clast imbrication (dashed lines) in parts (flow towards left); poorly sorted and clast supported; part of coarse-grained fluvial system. Scale bar 20cm.
Plio–Pleistocene, near Benidorm, SE Spain.

4.7   Sandy oligomict conglomerate in sandstone/pebbly sandstone sequence, poorly defined bedding, crude stratification in conglomerates, parallel and cross-lamination (bottom left) in sandstones; part of raised beach deposit with rounded limestone clasts mixed with locally derived dolerite clasts. Hammer 25cm. *Plio–Pleistocene, near Benidorm, SE Spain.*

4.8   Conglomerates, pebbly sandstones and sandstones; part of resedimented facies on subaqueous portion of fan delta; note slight unconformity (dashed line) probably due to active synsedimentary tectonics. *Miocene, Pohang basin, SE Korea.*

4.9   Polymict conglomerates, pebbly sandstones and sandstones; part of resedimented facies (coarse-grained turbidites) from deep-water slope apron/ submarine fan succession; distinct beds with gradational contacts, individual turbidites are coupled conglomerate–sandstone units (arrows); top to left. Width of view 1.5m. *Paleogene, central California, USA*

4.10 Oligomict conglomerate turbidite erosive into fine-grained silt-laminated mudstone turbidites; subparallel alignment of rounded tabular clasts showing little consistent imbrication; part of resedimented facies from deep-water slope apron/submarine fan succession. Hammer 25cm.
Paleogene, Matilija, California, USA.

4.11 Oligomict conglomerate, with crude bedding stratification, from subaerial part of alluvial fan/fan delta succession; clast-supported near base, matrix-supported in parts near top. Crude stratification (dashed lines) shows slight divergence of bedding from left to right (i.e. wedge geometry).
Carboniferous, Quebrada de las Lajas, NW Argentina.

4.12 ▶▲ Pebbly sandy mudstone/muddy poorly sorted boulder conglomerate; part of submarine debris flow deposit (debrite) with mixed hard and soft clasts. Note clast deformation. Lens cap 6cm.
Neogene, near Benidorm, SE Spain.

4.13 ▶ Volcaniclastic conglomerate/sandstone within tuffaceous marlstone succession. Dark clasts are scoriaceous (basic) volcanics; pale clasts are pumiceous (acidic) volcanics. Hammer 30cm.
Miocene, Miura Basin, south central Japan.

4.14 Pebbly mudstone (diamictite or tillite), bedding-plane view. Ancient glacial till deposit, very poorly sorted with clasts of mixed composition in a fine-grained matrix. Note several groove marks oriented top left to bottom right, and 'bulldozer' ruck marks in front of large clast 6cm left of the scale bar (in centimetres). *Late Carboniferous, Wynard foreshore, N Tasmania, Australia.*

4.15 Shale-clast conglomerate, part of resedimented deep-water massive sandstone succession. Bedding approximately horizontal. Hammer 45cm. *Miocene, Urbanian Basin, central Italy.*

4.16 Sandy muddy conglomerate, small part of over 50m thick Gordo Megabed, a tripartite resedimented unit (slide-debrite/slurry bed-turbidite) in small marginal marine deep-water basin succession, showing distinct bed-parallel alignment of tabular metamorphic clasts, and grain-size oscillation typical throughout megabed. Top to right, no overall grading evident. Hammer 45cm. *Miocene, Tabernas Basin, SE Spain.*

4.17 Chaotic oligomict breccia/conglomerate (limestone and volcanic clasts), part of sub-aqueous debris avalanche into small marginal marine basin. Hammer 45cm.

Late Triassic–Jurassic, Dolomites, N Italy.

4.18 Oligomict conglomerate with large rounded volcanic clasts and finer shelly debris in sandy matrix; part of debris ava-lanche deposit downflank from bioherm-capped submarine volcano. Width of view 60cm.

Miocene, Cabo de Gata, near Carboneras, SE Spain.

SANDSTONES

Giant sand dunes, Namib Desert, and diamond sand beaches, SW African coast.

Definition and range of types

SANDSTONES are all those medium-grained sedimentary rocks that comprise more than 50% sand-size (0.063–2mm) grains. They are also known as arenites, wackes and grey-wackes. Where the coarse-grained (clast) fraction increases they grade into pebbly sandstones and then sandy conglomerates; where the fine-grained (mud) fraction increases they grade from muddy sandstones into sandy mudstones.

Many different types of sandstone occur, that are generally recognized on the basis of composition (Table 5.1). Siliciclastic sandstones are the most important, being the standard detrital terrigenous sediment of sand grade (this chapter). Carbonate-rich sandstones are those with an abundance of calcareous grains and/or cement as well as siliciclastic material. These grade into sandy limestones with siliciclastic impurities, but are distinct from calcarenites, which are particulate limestones of sand grade (Chapter 6). Volcaniclastic sandstones or tuffs derive from explosive volcanic eruption and the breakdown of volcanic rocks (Chapter 14). Various other sandstones are named after a striking, but generally minor or accessory component, such as the greenish-coloured mineral glauconite in greensand, sparkling thin sheets of mica in micaceous sandstone, brown-black lignite in carbonaceous sandstone, and so on.

Principal sedimentary characteristics

See photographs and figures in the relevant sections of Chapter 3 as well as at the end of this chapter.

Bedding
- *Thickness*: very variable, from the thinnest beds and laminae to very thick-bedded massive sandstones in excess of 10m thick.
- *Shape*: regular tabular beds common; also irregular/lenticular beds with shape determined by bedform (e.g. dunes, sand waves), or by erosive basal structures (e.g. flutes, scours, channels).
- *Boundaries*: top and bottom boundaries typically sharp (± surface structures) where interbedded with mudrocks; indistinct/gradational in thick units of sandstone, and sandstone with conglomerate.

Structures
A wide range of primary and secondary structures are commonly observed in sandstones. These are very diagnostic of depositional process and environment as well as post-depositional modification. Note in particular: scale of cross-bedding (set height and

so on), paleocurrent indicators (lineation, sole marks, cross-lamination), sequence of structures in graded beds, bioturbation and trace fossils. Apparently structureless beds may reveal indistinct lamination, very slight grading or water-escape structures (dishes, pillars) on close inspection. Where possible, locate and examine the top and bottom surfaces of beds for characteristic bedding plane structures.

Texture
Sandstones have a mean size between 0.063–2.0mm, varying from very fine to very coarse grained and from very poor to very well sorted. They generally show good porosity and permeability characteristics on deposition but these become progressively occluded by compaction and cementation. Textural maturity in sandstones is a qualitative measure of the duration and energy-level of the transport-depositional history. More mature sandstones have well rounded grains and very low matrix content. Immature sandstones, by contrast, are poorly sorted, with angular grains and a high matrix content (*Fig. 3.29*). Grain fabric is more difficult to observe in the field than for conglomerates, but carefully oriented samples can be collected for later thin-sectioning where it is important to determine grain fabric.

Composition
Sandstones can include almost any pre-existing mineral or rock fragment but, due to removal of unstable grains by the very efficient agents of chemical and physical weathering, the typical composition is much more restricted. The principal components of most siliciclastic sandstones are quartz grains, feldspar grains and rock fragments. Other components can include micas, clay minerals, biogenic fragments (calcareous, siliceous and carbonaceous), and over 100 species of heavy minerals (specific gravity >2.9). Cement is precipitated around, between and within grains during diagenesis and can range from

5

Table 5.1 Principal types of sandstones

Major types	Subtypes	Nature and origin
Siliciclastic sandstone or arenite	Quartz sandstone or arenite (quartzarenite)	Compositionally mature sediment from the breakdown of older rocks, and the preservation of at least 90% quartz (typically by selective removal of other grains).
	Feldspathic sandstone or arenite (arkose, where feldspar >25%)	Compositionally submature–immature sediment from the breakdown of older rocks, and selective preservation of at least 10% feldspar during short distance/short duration transport and deposition.
	Lithic sandstone or arenite (litharenite)	Compositionally submature–immature sediment from the breakdown of older rocks, and selective preservation of at least 10% lithics during short distance/short duration transport and deposition.
	Muddy sandstone or arenite (wacke or greywacke; also matrix-rich sandstone)	Typically immature sediment from the breakdown of older rocks of many kinds, and containing fine-grained matrix (>15%) of detrital and/or diagenetic origin.
	Pebbly sandstone or arenite	Typically high energy deposit gradational to conglomerate.
Carbonate-rich sandstone	Calcareous sandstone	Sandstone with quartz and/or other detrital grains, together with 10–50% carbonate – either as grains (typically biogenics and ooids) and/or as carbonate cement.
	Sandy or quartzose limestone	Limestone (i.e. >50% carbonate) with up to 50% quartz or other detrital grains.
Volcaniclastic sandstone (tuff)	Vitric tuff	Glassy volcanic shards
	Lithic tuff	Grains of lava and pre-existing country rock.
	Crystal tuff	Euhedral/subhedral crystals (typically quartz and feldspar).
Other sandstones (hybrid sandstones)	Glauconitic sandstone	Common (>4%) glauconite grains (authigenic precipitate in sediment-starved marine shelf environments).
	Carbonaceous sandstone	Common (>4%) organic carbon grains and fragments (i.e. lignite, coal or brown sapropelic debris).
	Micaceous sandstone	Common (>4%) mica grains (typically detrital muscovite).
	Other subtypes also occur, such as gypsiferous sandstone, ferruginous sandstone	

being a minor to dominant component that serves to bind the rock together. The most common cements are quartz and calcite. Diagenetic hematite causes the red colouration of many sandstones, especially those of terrestrial environments.

The framework components of a sandstone are those of coarse silt and sand size (i.e. $>30\mu m$), whereas the matrix is the finer-grained ($<30\mu m$) interstitial material composed mainly of clay minerals and fine quartz/feldspar silt. Without laboratory petrographic analysis it is not possible to distinguish detrital from diagenetic matrix.

Compositional maturity in sandstones reflects the geology of the source area, the intensity of weathering and the length of the transport path. Grains that are mechanically and chemically more stable than others will resist breakdown to a greater extent (*Fig. 3.33*). Immature sandstones contain unstable rock fragments, plagioclase feldspars and mafic minerals, whereas increasing maturity

leads to progressively more quartz, with some potassic feldspars and stable rock fragments. Supermature sandstones comprise little other than quartz, chert and ultrastable heavy minerals (zircon, rutile, tourmaline).

Classification of sandstones

AMID the confusion of classification schemes that exist for sandstones, most of which are based on the relative proportions of different components, one of the most accepted schemes is given in *Fig. 5.1*. The three main framework components give rise to the terms quartz sandstone or arenite (also quartz-arenite), feldspathic sandstone or arenite (also known as arkose where the feldspar content exceeds 25%, and subarkose where it is 10–25%), and lithic sandstone or arenite (also litharenite). Muddy or argillaceous sandstones (also matrix-rich sandstones, wackes, greywackes) are those with a matrix

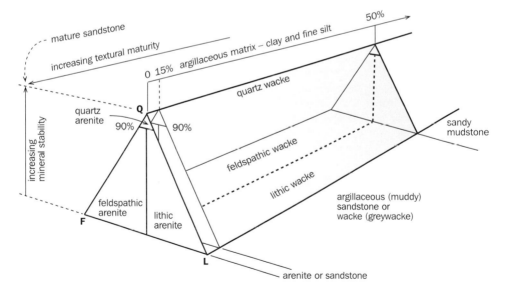

5.1 **Classification of sandstones based on grain type and matrix content. Note that Q = quartz, chert, quartzite grains, F = feldspars, L = lithic (rock) fragments. Arenites are those sandstones with little matrix (mud) content – here taken as <15%. Wackes are those sandstones with >15% matrix.** (*Modified after Pettijohn et al 1987*).

content (i.e. <30μm material) that exceeds 15%. Whereas most early classifications extend the wacke range to <75% matrix, I prefer to use 50% as the top end of the range. Any more matrix/mud than this and the sediment becomes a mudrock. Modifiers, such as micaceous, calcareous, carbonaceous, glauconitic and ferruginous can be added as prefixes where appropriate. There is no generally accepted proportion of the minor component before such prefixes are used – I suggest >4% in Table 5.1, although others will use >2% (especially for mica).

These main sandstone types can generally be recognized in the field by careful use of a hand-lens to identify minerals. (This is essential!) Percentage estimation charts can then be used to estimate the proportions of various constituents present. Colour is also a guide. Quartz sandstones are pale grey to whitish, except where coloured red by hematite in subaerial environments. Feldspathic sandstones are pinkish or red due both to feldspars and hematite; they may also be speckled with white 'grains' where the feldspars have altered to clay minerals (especially kaolinite). Lithic sandstones have more varied colours depending on the principal rock fragments present. Muddy sandstones, including the archetypal greywackes, are generally dark grey in colour due to the higher matrix (clay) content. Any sandstone can take on a yellow, orange or brownish-red hue where even very small quantities of iron-rich minerals are present. These are readily leached into the groundwater by weathering and then precipitate out as oxides and hydroxides.

Occurrence

SANDSTONES are one of the more common sediment types and can be found in nearly all environments from the most proximal alluvial fan to deepest marine basin plain. Quartz sandstones are most typical of moderate to high-energy shallow marine environments,

---

FIELD TECHNIQUES

Don't be confused by weathering colours and patterns (e.g. liesegang rings, dendrites). Find a fresh surface, or carefully knock off an edge or corner with your hammer, and examine with a hand lens. Quartz grains are milky to clear, glassy, with no cleavage and fresh conchoidal fractures. Feldspars are generally white or pinkish, with cleavage planes and traces on fracture surfaces. They are often partly altered to powdery clay minerals or dissolved out as holes. Lithic grains are composite and varied in nature.
See Tables 3.1–3.3 for further details.

---

their correlative deep marine deposits, eolian sand seas of deserts, and sand dunes along the seashore. They also form by leaching and dissolution of all unstable grains. Ganisters can be formed in this way as the seat-earth with rootlet traces underlying coal seams. Feldspathic and lithic sandstones are common in alluvial fan, fluvial, and some lacustrine environments, as well as deltaic and deep water environments, especially in tectonically active regions. Muddy sandstones (greywackes) were once thought most typical of turbidites in deep water environments. However, it is now recognized that they also occur in many other settings (e.g. flood plain, deltaic, and outer shelf) where the transporting energy is low to moderate, whereas turbidites can be of all compositional types.

Sedimentary structures can yield much information on processes of sandstone deposition, textural characteristics are especially important when considering reservoir properties, and sandstone composition will influence its potential use as an industrial mineral (in glass making or for building sand, for example). Bed thickness, geometry and the 3D architecture of sand bodies is vital for economic studies related to the oil, coal, and water industries.

Field photographs

5.1 Interbedded turbidite sandstones (hard) and mudstones (soft); each sandstone–mudstone couplet is a single turbidite; note sharp bases and tops to individual sandstone layers typical of ancient turbidites. Width of view 4m.
Eocene, near Annot, SE France.

5.2 Immature lithic feldspathic sandstone (or lithic arkose). Part of a very thick-bedded deepwater turbidite succession. Some of the beds are 1–10m thick, with no grading and no structures (structureless) – these are known as deepwater massive sandstones, probably deposited from high-concentration turbidity currents. Some show subtle water escape features (dishes and pipes) or, as in this case, zones of indistinct parallel lamination. Mobile phone beneath root system 10cm.
Pliocene, Umegase, Boso Peninsula, Japan

5.3 Medium to thick-bedded calcareous lithic sandstones with thin mudstone partings; scoured–loaded bases and parallel to wavy internal lamination; part of channel-fill sequence in subaqueous fan-delta front turbidite succession.
Width of view 1.3m.
Miocene, Tokni, S Cyprus.

5.4 Coarse, lithic, muddy feldspathic sandstone (or lithic greywacke–arkose) and pebbly sandstone, erosive into fine laminated sandstone. Note reverse grading (reverse arrow, R) apparent through lower 40cm of coarse sandstone turbidite, which is otherwise structureless.
Hammer 30cm.
Triassic, near La Serena, Chile.

5.5   Lithic sandstone and pebbly sandstone with crude parallel stratification, typical of resedimented facies on subaqueous portion of fan-delta succession.
Miocene, Pohang Basin, SE Korea.

5.6   Parallel and cross-laminated quartz sandstone with climbing ripples (dashed lines) and isolated ripples; brown-coloured layers are finer-grained (muddy) material; turbidite succession, flow to right. Width of view 20cm.
Eocene, near Annot, SE France.

5.7   Quartz sandstone, laminated and cross-laminated (planar tabular cross-sets), with vertical burrows increasing in abundance towards top of section; shallow marine estuarine succession. Width of view 1.1m.
Pliocene, Sorbas Basin, SE Spain.

5.8 Quartz sandstone showing large-scale hummocky cross-stratification (above hammer) within parallel-laminated section; shallow-marine storm-influenced succession. Hammer 25cm. *Carboniferous, Quebrada de las Lajas, NW Argentina.*

5.9 Muddy sandstone (greywacke); part of thick-bedded turbidite showing parallel and large-scale wavy lamination – such swaley to hummocky cross-stratification is unusual in turbidites. Width of view 60cm. *Cambrian, Bray Head, Eire.*

5.10 Bioclastic (shell-rich) lithic sandstone (dark lithic clasts more evident in lower part of view), crudely laminated and cross-laminated, very coarse-grained to pebbly in parts. Lens cap 6cm. *Pliocene, Pissouri Basin, S Cyprus.*

5.11 Calcareous lithic sandstone with dark volcanic debris and white bioclasts, parallel and cross-lamination with low-angle erosional (reactivation) surface near top (dashed line); lower foreshore succession. Hammer 25cm.
Pliocene, Pissouri Basin, S Cyprus.

5.12 Muddy glauconitic sandstone unconformably overlain by flint conglomerate; greenish coloured glauconitic unit is a highly bioturbated, shallow-marine deposit. The iron in glauconite is typically oxidized during weathering to limonite (brownish colour). Note that vertical marks are an artefact of scraping clean the surface. Hammer 45cm.
Eocene beneath Pleistocene, Lee-on-the-Solent, S England

5.13 Feldspathic sandstone (arkose) with pale green–beige reduction zones and cross-lamination; typical of red bed fluvial succession. Hammer 45cm.
Permo–Triassic, Nottingham, central England.

5.14 Complex cross-laminated quartz sandstone with rare pebble horizons; eolian back-beach dune sands with storm wash-over pebbles.
Notebook 20 cm high.
Devonian–Carboniferous, Lladwyn Island, Anglesey, UK.

5.15 Feldspathic muddy sandstone (feldspathic greywacke); bedding-plane view showing large fossil-wood impressions in deltaic succession.
Hammer 25cm.
Carboniferous, Ashover, Derbyshire, UK.

5.16 Feldspar-rich sandstone with some bioclastic debris (white); parallel lamination with low-angle scour (reactivation surface, dashed line) and minor fault (right); shallow-marine succession, lower foreshore.
Hammer 25cm.
Silurian, near Canberra, ACT, Australia.

5.17 Lithic feldspathic sandstone (arkose) with shale clasts. Very thick-bedded and structureless sandstone turbidite (deep-water massive sand facies); common shale clasts near base probably due to complete erosion and break up of underlying mudstone. Prominent joint with weathering and discolouration to right. Hammer 25cm. *Eocene, Matilija, California, USA.*

5.18 Lithic muddy sandstone (lithic greywacke) turbidite bed (arrow), showing erosive base, normal grading and standard Bouma sequence of structures (as marked A–E). Width of view 60cm. *Ordovician, Southern Uplands, Scotland.*

MUDROCKS

Cyclic bedding in deep-water mudrocks, Devonian Ohio Shale, USA. *Photo by Paul Potter.*

Definition and range of types

MUDROCKS are the most abundant of all lithologies, making up at least 50% of sedimentary rocks in the stratigraphic record. However, because of their grain size, and their subjectivity to weathering and landslides, they present more of a challenge in the field and have been relatively understudied. This trend is fast changing.

Mudrock is the preferred general term for the large group of fine-grained siliciclastic rocks, composed mainly of particles <63μm in size. Common synonyms in use by other authors include shale, mudstone and lutite, although these have slightly more restricted meanings in this classification. Mudstone (or lutite) is blocky and non-fissile, whereas shale is fissile. Both may be laminated. Mudrocks can be further subdivided on the basis of grain size (siltstone and claystone), sediment fabric (mudstone and shale) and composition (see below). Because metamorphic grade mudrocks are commonly better preserved and more easily studied than their unmetamorphosed equivalents, the terms argillite, slate and phyllite are also shown in Table 6.1 of mudrock genetic types.

Mixed biogenic–siliciclastic mudrocks are very common lithologies and the term marl (or marlstone) has long been used for calcareous mudstones and muddy chalks or limestones. More recently, the terms sarl, for siliceous biogenic mudstones, and smarl, for mixed composition sediments, have been introduced. Black shale is a widely used term for generally fine-grained, organic-rich sediments; these grade into oil shales with increasing organic carbon content. Volcaniclastic mudstones are known as fine-grained tuff or volcanic dust deposits (Chapter 14).

Principal sedimentary characteristics

See photographs and figures in the relevant sections of Chapter 3 as well as at the end of this chapter.

Bedding
- *Thickness*: very variable, from apparently very thick unbedded units to very finely laminated sections.
- *Shape*: includes regular, tabular, extensive units (thin and thick), but also abrupt terminations due to erosive cut-out; individual laminae show wide range of planar, lenticular and contorted shapes.
- *Boundaries*: sharp or gradational; top of bed subject to erosional and bioturbational markings; compaction and diagenesis typically accentuates the sharpness of mudrock boundaries so that they appear more abrupt than when first deposited.

Structure
The effects of weathering and cleavage can seriously obscure structures in mudrocks but, in other cases, close inspection will be very rewarding. Microstructures that are commonly observed in thin silt laminae within finer-grained mudstone include: parallel, cross- and convolute lamination; grading and graded-laminated units; scours, loads and water-escape features. Very close inspection of the outcrop is necessary, even use of a hand lens to determine whether lamination is truly parallel or micro-cross-laminated. Bedding plane surfaces may reveal current lineation, alignment of fossils (e.g. graptolites), micro-scours and micro-flutes. Note paleocurrent directions. Distinct burrow traces and irregular bioturbational mottling, as well as diagenetic nodules or concretions (calcite,

dolomite, siderite, pyrite), are common throughout many mudrock successions.

Fabric
Fissile lamination (or fissility) is an irregular, subparallel, fine-scale parting fabric particularly common in black shales and in many compacted hemipelagic mudrocks. It is probably mainly due to the bed-parallel alignment or flat-packing of platy grains aided, in the case of black shales, by the presence of organic matter. Paper shales have very fine fissility (<1mm partings); platy shales have coarser fissility (>1mm partings). Normal parallel lamination is more regular and clearly picked out by variation in grain size, composition and/or colour. Splitting along this lamination can lead to flaggy (5–10mm partings) and slabby (>10mm partings) mudrocks.

Texture
Mudrocks have a majority of grains in the silt and clay grades (i.e. <63μm). Within this caveat there are a wide range of types including coarse pebbly and sandy mudrocks, granular siltstones, fine cohesive mudstones and claystones, and fissile shales. Only the coarsest silt sizes and above can be resolved in the field with the aid of hand lens. Otherwise, for semi-consolidated rocks, silts have a gritty feel between the fingers or when chewed, whereas clays are smooth and pasty. Colours can also be indicative of grain size. For siliciclastic mudrocks, silt laminae are generally light coloured, whereas finer-grained mudstones or claystones are dark coloured.

Microfossils are typically some of the coarser particles present and are visible with careful use of the hand lens. Sorting and grain morphology can only be determined by laboratory analysis.

Composition
Clay minerals, fine micas, quartz and feldspars are the most abundant components of siliciclastic mudrocks. Fine-grained detrital and biogenic carbonate (and silica) are

Table 6.1 Principal types of mudstones

Major types	Subtypes	Nature and origin
Siliciclastic mudrock (terrigenous or clastic mudrock)	Claystone (clay-grade dominant) Siltstone (silt-grade dominant	Finer-grained and coarse-grained mudrocks (respectively), the former with more clay minerals, the latter with more quartz silt.
	Mudstone (non-fissile) Shale (fissile)	Subtypes characterized by fabric; shales have a tendency to split into thin sheets along a strongly aligned fabric; mudstones are non-fissile with a blocky or massive fabric.
	Argillite Slate Phyllite (mica schist)	Subtypes based on increasing metamorphic grade: argillite (low) to phyllite (high); structures typically well preserved in low and medium grades.
Mixed siliciclastic–biogenic mudrock	Marl (stone)	Biogenic calcareous mudstone to muddy chalk or limestone. (Not always biogenic.)
	Sarl (stone)	Biogenic siliceous mudstone to impure chert.
	Smarl (stone)	Mixed biogenic calcareous and siliceous mudstone.
Pebbly mudrock/mudstone (diamictite)	Glacial diamictite	Chaotic, massive, unbedded unit, with features indicating glaciomarine, glaciolacustrine, or subglacial moraine deposition.
	Debrite (olistostrome, lahar)	Chaotic to poorly structured, massive to bedded unit, with features indicative of subaerial or subaqueous debris flow deposition.
	Melange (include mud-rich and gravel-rich varieties)	Chaotic, massive, unbedded unit, with exotic and outsize clasts in fine-grained matrix, +/- pervasive tectonic shear. Associated with major faulting, ophiolite emplacement, mud volcanoes, etc.
Other types	Black shale	Organic-carbon-rich (>1% TOC) fine-grained sediment (generally mudrock).
	Red clay (stone)	Very deep open ocean sediment accumulating below carbonate compensation depth.
	Red mudstone	Common continental (fluvial) deposit (of the 'red bed' suite).
	Variegated mudstone	Multicoloured mottled mudrock, most typical of continental and shallow-marine deposits.
	Volcaniclastic mudstone (fine-grained tuff)	See Ch. 14

equally important in the marl–sarl family. A variety of other grains occur in minor quantities or, in some cases, more abundantly. These include volcanic ash, zeolites, iron oxides and sulfides, heavy minerals, sulfates, and organic matter (kerogen). Microfossils, macrofossils and diagenetic nodules may also be common. There is a close relationship between grain size and certain components of mudrocks (*Fig. 6.1*). The principal constituents of mudrocks is given in Tables 3.1 and 3.3 (Chapter 3). For the most part, however, mudrock composition must be determined in the laboratory by x-ray diffraction, electron microscopy or a variety of geochemical techniques.

Colour can be some indication of composition in mudrocks (see page 121). They occur in a variety of colours ranging through red, purple, brown, yellow, green and grey to black. Various shades of light, medium and dark grey are the most common. The darker colours are generally due to preservation of organic carbon, becoming very dark grey or black in black shales and oil shales. The paler colours are either due to a siltier texture (with

significant amounts of silt-sized quartz and feldspar), or to a greater biogenic content (especially calcareous material). The red to green spectrum generally depends on the oxidation state of the iron present – completely oxidized terrestrial mudrocks are typically red, with more reduced zones being greenish in colour.

Classification of mudrocks

TWO PRINCIPAL schemes are used in the classification of mudrocks but neither is very easy to apply in the field! The first is based on relative proportions of the three grain size components: clay, silt and sand (*Fig. 6.2*, front panel). The second is based on relative proportions of the three mineral components: mud, biogenic carbonate and biogenic silica (*Fig. 6.3*, front panel). The most readily applied field terms are mudstone (non-fissile), shale (fissile), siltstone (granular but not sand grade), and marlstone (calcareous). In order to include the very important and common mudrock types, pebbly mudstones and black shales, each of the classification schemes in *Fig 6.2* has been modified by adding a further axis.

Black shale is the general term applied to all those dark-coloured, fine-grained sediments (including mudrocks, marls and micrites) with relatively high amounts of organic carbon. Since organic carbon is rarely preserved in sediments, in more than trace amounts, then >1% is taken as the qualifying figure for black shales. These sediments are potential source rocks for hydrocarbons, depending on the type and amount of organic matter present. Good source rocks may contain from a few percent to 20% organic carbon – especially that derived from plankton and bacteria. However, organic carbon is mainly present as kerogen, a complex

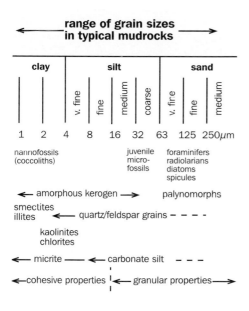

6.1 Relationship of grain size to the principal components of mudrocks.

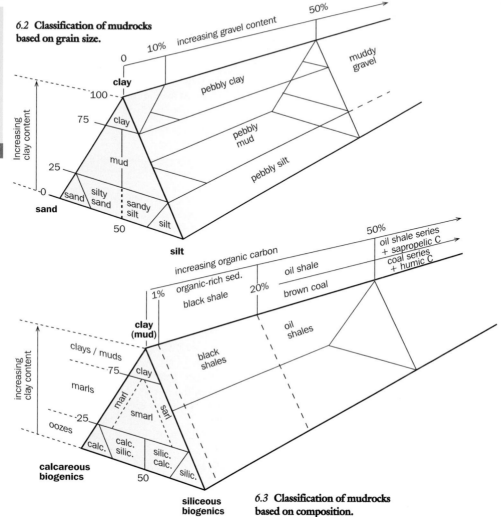

6.2 Classification of mudrocks based on grain size.

6.3 Classification of mudrocks based on composition.

geopolymer of high molecular weight formed from the progressive diagenesis of organic matter during burial and heating. It is too fine and dispersed to be visible in the field, even using a hand lens.

Oil shale is a very organic-rich black shale, having in excess of 20% organic carbon, mostly as kerogen and possibly with some bitumen present. The kerogen in many oil shales has been derived from algal material, while any liquid heavy oil may be found oozing from cracks and joints in shale. Oil

shales can act as hydrocarbon source rocks, but also provide a direct source of fossil fuel. Oil can be extracted by direct heating (distillation) of the shale, yielding 10–150 gallons per ton of rock. Oil shales are soft and even malleable; they can be cut with a knife into thin shavings that curl like wood shavings.

Pebbly mudstones and bouldery mudstones are typical chaotic sedimentary facies formed as glacial diamictites, subaerial and subaqueous debrites (olistostromes), and tectonic melanges (see page 68–70).

Occurrence

MUDROCKS are ubiquitous, especially in quieter, low-energy depositional environments, including: alluvial–fluvial flood plains; lacustrine, lagoonal, and estuarine environments; interdistributary delta areas and pro-delta slopes; outer shelf and all parts of the deep-water slope to basin-plain environment. The depositional environment is best determined from a combination of sedimentary structures, trace fossils, micro and macro-fossils, and associated facies. Black shales and oil shales require special conditions, generally involving excess supply of organic matter (plankton blooms, upwelling regions, terrestrial run-off) and/or anoxic conditions that favour preservation of organics. Rapid burial and fine sediment size also aid preservation. Pebbly mudstones may result from debris flow and glacial processes, either on land or at sea.

> **FIELD TECHNIQUES**
> Don't give up too easily on mudrocks where they are badly weathered or poorly exposed. Hunt around for fresher outcrops; or even resort to digging, trenching and cleaning the surfaces. There is often a wealth of hidden information to be exposed. Sample routinely for microfossil biostratigraphy.

Field photographs

6.1 Red calcareous mudstone over limestone–marlstone succession. Reclining figure in mid distance is lying on the K–T boundary.
Paleocene over Cretaceous, near Urbino, central Italy.

6.2 Thick succession of mudstone (dark) and siltstone (light) turbidites and hemipelagites. Width of view 80m.
Eocene, Mugu Point, California, USA.

6.3 Thin horizons of red calcareous mudstone within micrite succession, formed mainly through dissolution of carbonate fraction from original marls; note side view of large *Zoophycos* (trace fossil) mound left of lens cap (dashed line). Lens cap 6cm.
Cretaceous, near Urbino, central Italy.

6.4 Thin horizons of red mudstone/marlstone interbedded with micrites. Width of view 30cm.
Silurian, Murrumbidgee, ACT, Australia.

6.5   Laminated siltstone–mudstone turbidites. Siltstone laminae (pale coloured), mudstone laminae (dark grey). Fallen blocks in foreground; inclined laminated unit to top left is *in-situ*. Lens cap 6cm.
Carboniferous, Rio Tinto, S Spain.

6.6   Laminated and thin-bedded volcaniclastic turbidites. Siltstone laminae (pale coloured), mudstone laminae (dark grey). Thin bed of contorted silt laminae (mid-view) is the result of dewatering disruption, probably due to seismic shock. Note low angle normal fault (dashed) near top of view. Hammer 25cm.
Cretaceous, St Croix, US Virgin Islands.

6.7   Organic-carbon-rich finely laminated mudstone with glacial dropstone (approximately 4cm diameter).
Photo by Jan Zalasiewicz.
Ordovician, central Wales, UK.

6.8 Pebbly mudstone with dispersed irregular sized clasts in dominant matrix support of very poorly sorted sandy mudstone. Glacial diamictite.
Lens cap 6cm.
Carboniferous, West Yunnan, China.

6.9 Muddy and silty laminated/bioturbated contourites (probable shallow-water contourites); note coarsening-up to fining-up sequence (arrows) typical of contourites, and local erosion surface towards top.
Hammer 30cm.
Triassic–Jurassic, Los Molles, central Chile.

6.10 Volcaniclastic calcareous muddy contourite; note ill-defined layering, streaky concentrations of coarser volcaniclastic sands, and bioturbation throughout. Lens cap 6cm. *Miocene, Miura Basin, near Tokyo, Japan.*

6.11 Volcaniclastic calcareous hemipelagic mudstone with sideritic concretions and bioturbation throughout. Hammer 25cm. *Miocene, Miura Basin, near Tokyo, Japan.*

6.12 Silt-laminated mudstone unit (viewed by geologist and dog). Lower part of unit is more silt-rich (lenticular to flaser lamination); upper part is more mud-rich (wavy to lenticular lamination). Probable shallow water, estuarine depositional environment. *Triassic, Los Molles, west central Chile.*

6.13 Dark-coloured fissile mudrock (shale) with large fractured ironstone concretion. Note thickening of laminae into concretion indicates formation was partly pre-compaction. Probable estuarine depositional environment. Coin 2.5cm. Photo by Paul Potter.
Carboniferous (Pennsylvanian), Caseyville Formation, S Illinois, USA.

6.14 Thin-bedded turbidite siltstones/sandstones (pale brown) and silt-laminated mudstone turbidites (dark grey); lacustrine setting with volcaniclastic input.
Triassic, Puquen, west central Chile.

6.15 Silt-laminated mudstone turbidites; mainly parallel laminae, some long-wavelength low-amplitude ripples (LAR) and isolated fading ripples (FR) at bases of partial Stow sequences (arrows).
Lens cap 6cm.
Carboniferous, Quebrada de las Lajas, NW Argentina.

6.16 Laminated lagoonal silt-stone and mudstone with dark organic-rich layers; brackish-water to shallow-marine setting. Width of view 40cm.
Pliocene, Sorbus Basin, SE Spain.

6.17 Lacustrine silty mudstone with duricrust-coated rootlets and small, whitish nodules (probable gypsum caliche). Hammer 25cm.
Plio–Pleistocene, Goyder, Lake Eyre, Australia.

6.18 Thick, poorly bedded loess deposit–windblown siltstone. This is part of the Peorian loess, the latest and most widespread of many loess deposits in the Midwestern USA. It is thickest near the Mississippi River from which it was derived by defla-tion of a meltwater outwash floodplain. Section height 5m. Photo by Paul Potter.
Quaternary (Wisconsin), Vicksburg, Mississippi, USA.

6.19 Variegated mudstones and marlstones (purple, red, green, and yellowish colours); now highly weathered and disturbed by recent landslides. Brackish water lagoonal environment. Width of view 60cm.
Cretaceous, Compton Bay, Isle of Wight, UK.

6.20 Fine sandstone/siltstone to mudstone turbidites (arrows). Note typical weathering of softer, more fissile mudstones obscures primary features. Hammer 25cm.
Cretaceous, central Honshu, Japan.

6.21 Part of very thick succession of reddish-coloured, silt-laminated mudstone turbidites (partial Stow sequences, arrows). Siltstone is paler coloured. Red colouration probably due to presence of highly oxidized iron (FeIII) from continental weathering of Precambrian hinterland. Width of view 20cm.
Late Precambrian, Hallett Cove, S Australia.

6.22 Interbed of dark-coloured mudstone with contorted siltstone interval (paler colour, arrow), within deep-water sandstone–mudstone turbidite succession. All now show low-grade metamorphism. Contorted siltstone may be due to rapid deposition from unstable turbidity current, or to subsequent liquefaction caused by seismic tremor.
Width of view 65cm.
Late Precambrian, Langkawi Island, Malaysia.

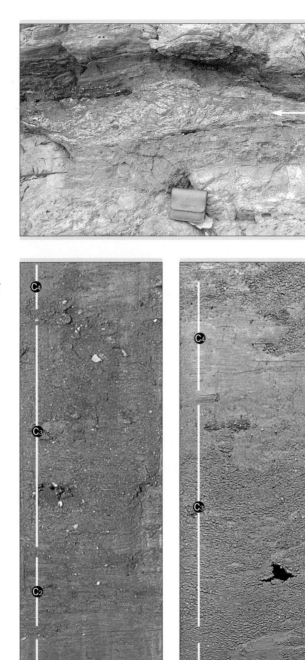

6.23 Core sections split lengthways through contourite drift sediment, showing parts of standard contourite sequence. From base to top: C1 mud, C2 mottled silty mud, C3 muddy sand, C4 mottled silty mud. Note white bioclastic (shell) debris in near section, division C3. Also note bioturbation throughout and partly disrupted, discontinuous lamination with some sharp contacts.
Width of core 8cm.
Pleistocene, Faro Drift, Gulf of Cadiz.

CARBONATE ROCKS

Waulsortian mound, Fort Payne Limestone Formation, Kentucky, USA. *Photo by Paul Potter.*

Definition and range of types

CARBONATE ROCKS make up about one-fifth of all sedimentary rocks in the stratigraphic record. They are defined as comprising over 50% carbonate minerals, dominantly calcite and dolomite in ancient rocks, and calcite (including high and low-Mg varieties) and aragonite in modern sediments. Aragonite is commonly replaced by calcite during early

diagenesis (only rarely being preserved in some fine-grained impermeable limestones) and Mg is generally lost from high-Mg calcite. The other carbonate minerals are only rarely present in limestones and dolomite.

The major groups of carbonate rocks, therefore, are limestones and dolomites, including transitional or partly dolomitized limestones (Table 7.1). All carbonate sediments are more or less affected during diagenesis by significant dissolution, replacement of aragonite, and so on, whereas dolomites are commonly wholly formed by early to late-stage diagenetic replacement of limestone.

Many limestones are directly analogous to siliciclastic rocks, being formed by the transport and deposition of carbonate particles: gravel-sized being known as calcirudites, sand-sized as calcarenites, and mud-sized as calcilutites. However, the carbonate sand and mud particles are typically formed by biogenic or chemical precipitation, rather than weathering and erosion.

Other limestones form *in situ* by the growth of carbonate skeletons in a reef framework, or by the trapping and binding of sediment in microbial mats. Reef limestones generally have a structureless, unbedded appearance, with the skeletal material of colonial organisms as a framework for further growth, and cavities filled by a variety of carbonate debris and cement. The term 'patch reef' refers to a small localized buildup, typically circular to oval in plan; whereas a barrier reef is a larger, elongate structure, typically associated with lagoonal limestones to the landward and more extensive reef talus on the seaward margin. The terms 'bioherm' and 'biostrome' can be used as synonyms with patch and barrier reefs, respectively, but they are also used where the buildup lacks any clear skeletal framework.

Mixed carbonate–siliciclastic rocks are also common, with all gradations between carbonate-rich siliciclastics (e.g. calcareous sandstones) and siliciclastic (or impure) limestones (e.g. sandy limestones). Marl or marlstone is a good field term for a true hybrid of mudstone and fine-grained limestone (calcilutite or micrite). When it contains over 50% siliciclastic mud it is classed as a mudrock, whereas with over 50% carbonate it is a limestone. In the field, it is impossible to precisely determine component percentages, so the general term 'marl' (omitting the '-stone') is widely used. (See also Chapter 6, *Fig. 6.3*).

Principal sedimentary characteristics

See photographs and figures in the relevant sections of Chapter 3 as well as at the end of this chapter.

Bedding

- *Thickness*: very variable, from extremely thick unbedded units to very fine microbial-laminated sections.
- *Cyclic bedding*: regular alternation of limestone–marl, limestone–chert (or similar facies) is very common.
- *Shape*: includes regular, tabular, extensive units (thin and thick), as well as subparallel/lenticular bedding. This latter, more irregular bedding is the common result of post-depositional dissolution. Reef limestones are structureless and unbedded, typically in contrast to associated well-bedded limestones. The reef talus slopes are also structureless, and irregular to wedge-shaped.
- *Boundaries*: sharp or gradational; dissolution layers (marls/clays), dissolution seams and stylolites are very common; original bedding surfaces not always apparent; compaction and diagenesis in some cases accentuates the sharpness of limestone boundaries so that they appear more abrupt than when first deposited.

Structure

Carbonate rocks display the full range of primary and secondary sedimentary structures found in their siliciclastic equivalents. Abundant lamination and cross-lamination, including large-scale cross-bedding, are associated with shallow-marine carbonates with tidal and shelf influence. Graded bedding, sequences of structures (e.g. Bouma or Stow sequences), loads and scours, and structureless or chaotic units are associated with downslope resedimentation into deeper water by turbidity currents, debris flows, slides and slumps.

Cyclic bedding (e.g. marl–limestone) and extensive bioturbation are typical of pelagic

Table 7.1 Principal types of carbonate rocks

Major types	Subtypes	Nature and origin
Limestones CaCO₃ dominant	Subdivided according to: **1. Grain size** (cf. siliciclastics) Calcirudite >2mm Calcarenite 0.063–2mm Calcilutite <0.063mm **2. Main constituents** (Folk) Oosparite, oomicrite Pelsparite, pelmicrite Biosparite, biomicrite Intrasparite, intramicrite Biolithite, dismicrite **3. Textural features** (Dunham) Grainstone, packstone, wackestone, mudstone, boundstone, floatstone, rudstone, bafflestone, bindstone, framestone	Carbonate particles formed by primary chemical precipitation, by biogenic secretion, as fragmented remains of carbonate skeletons, and by erosion of pre-existing carbonate rocks.
Dolomites (Ca Mg (CO₃)2)	Subdivided according to the degree of dolomitization: <10% dolomite = limestone 10–50% dolomite = dolomitic limestone 59–90% dolomite = calcic dolomite >90% dolomite = dolomite	Most dolomites formed by partial to complete replacement of limestone; can occur penecontemporaneously in evaporitic conditions, but most form during shallow- to deep-burial diagenesis.
Carbonate–siliciclastic mixed rocks	Polymict (carbonate) conglomerates–impure calcirudites Sandy limestones–impure calcarenites Marlstones/smarlstones	Generally detrital carbonate particles and cement mixed with siliciclastic material in varying proportions. *See also* Mudrocks (Ch. 6)

and hemipelagic successions. Grain-size cyclicity, associated with bioturbation and minor, indistinct primary structures, is common in contouritic sequences. Structures that are more common to carbonates than siliciclastic sediments include cavity structures, hardgrounds and tepee structures, paleokarst surfaces, stromatolites, and framework reef structures (see page 102).

Fabric

The arrangement of grains and larger clasts is as important an aspect of carbonate rocks as it is for siliciclastics. Carbonates are typically clast-, grain- or matrix-supported; cement may be more dominant than in most siliciclastic rocks and framework-support is typical of *in situ* growth (e.g. reefs and bioherms). Bed-parallel grain alignment and clast imbrication are also typical fabric types.

Texture

- *Grain size*: carbonate rocks span the whole range of grain sizes including mud grade (calcilutites), sand grade (calcarenites), and gravel grade (calcirudites). The subdivisions of these classes are the same as used as for the equivalent siliciclastic sediment. Where possible, mean grain size, maximum grain/clast size, sorting, and grain shape should be estimated by use of comparator charts (see end pages).
- *Depositional texture*: one of the principal classification schemes in use for carbonates is based on the relative proportion of carbonate mud present and on fabric type (grain or mud support). This was called depositional texture by Dunham (1962) (see below and *Fig. 7.2*).

Composition

The chief components of most modern carbonate sediments (Table 3.2) are skeletal grains (bioclasts or fossils), ooids and other sub-rounded grains, intraclasts, carbonate mud or micrite and, in bioherms and ancient rocks, cement (e.g. sparite).

Skeletal grain type depends on environmental factors (salinity, temperature, depth) as well as on age and stage of evolution. They occur in Phanerozoic but not Proterozoic limestones. The main organisms that contribute skeletal debris include molluscs, brachiopods, corals, echinoderms, bryozoans, calcareous algae, and foraminifera. Other groups may be locally important – sponges, stromatoporoids, crustaceans (especially ostracods), annelids, and tentaculitids. Identifying the types of organisms present yields important information for both biostratigraphy and paleoecology (see pages 116–120). Many corals, stromatolites, bryozoans and algae are bioherm-building organisms that prefer shallow-marine environments. Thick-shelled molluscs and brachiopods are well adapted to high-energy nearshore waters, whereas free-swimming and planktonic organisms (including many microfossils that can be difficult to recognize in the field) tend to accumulate preferentially in quieter open waters. It is also important to note the state of fragmentation and whether or not the skeletal material is in growth position (framework, baffle or encrustation).

Ooids are spherical to subspherical, sand-size grains (typically 0.2–0.5mm diameter). Where the diameter is >2mm they are called pisoids. Both consist of concentric carbonate coatings around a nucleus (e.g. bioclast or quartz grain). Most modern ooids are aragonite, although this changes to calcite during diagenesis, and some ancient ooids were originally calcite. They are formed by direct precipitation of carbonate from shallow-marine, warm, and agitated waters, quite possibly using a microbial intermediary to help set up the ideal local chemical conditions that favour precipitation. More irregularly shaped sub-spherical particles (similar in size to and often confused with pisoids) are known as oncoids. These are, in fact, due to microbial precipitation of carbonate as the growing biogenic particles are washed to and fro in shallow agitated waters.

Peloids are subspherical to ellipsoidal grains (typically <1mm long), composed of fine-grained calcite (micrite) and having no internal structure. They mostly originate as faecal pellets, especially from organisms such as gastropods, crustaceans and polychaetes, and occur in limestones of more protected environments – lagoons and tidal flats, for example. More irregularly shaped amorphous grains can be formed by microbial replacement of skeletal particles by micrite.

Intraclasts are particles and clasts of reworked carbonate sediment (lithified or partly lithified), commonly from sand-sized up to several cm in length. They form from the desiccation of tidal-flat muds, erosion of tidal, lagoonal and shelf muds by storms, and of deeper marine slope and basinal carbonates by the passage of turbidity currents.

Micrite is the term given to fine-grained carbonate material (microcrystalline calcite, <4μm in diameter) that forms the matrix of many limestones and is the main constituent of fine-grained limestones. Its modern precursor is carbonate (or lime) mud that mainly forms through the disintegration of calcareous algae and other skeletal material, and in part through direct inorganic precipitation. Micrite can also form directly as a cement rather than a matrix.

Carbonate cement types in limestones include: sparite, a clear, equant, pore-filling calcite; microsparite, a calcite cement with smaller crystal size (5–15μm); fibrous calcite, as a coating on grains and fossils and a lining on cavity walls; and micrite (a micro-crystalline calcite).

Colour

Grain size and impurities are the key factors that influence the colour of carbonate rocks (see also page 121). Very pure carbonates, both coarse and fine-grained, are very light grey to white in colour. In less pure limestones, the darker grey colours are associated with finer grain size. The nature of the impurity further determines colour: organic-rich limestones are dark grey to black, whereas increasing iron content makes for pinkish to reddish colours. The iron content of many dolomites typically give these rocks a pinkish to orange-brownish hue.

principal allochems in limestone	limestone types			
	cemented by sparite		with a micritic matrix	
skeletal grains (bioclasts)	biosparite		biomicrite	
ooids	oosparite		oomicrite	
peloids	pelsparite		pelmicrite	
intraclasts	intrasparite		intramicrite	
limestone formed *in situ*	biolithite		fenestral limestone – dismicrite	

7.1 **Classification of limestones based on composition.** *Modified after Folk 1962.*

Classification of limestones

THREE schemes are currently used for classifying limestones.

1. *Grain size*. A simple but useful scheme divides limestones on the basis of mean grain size into calcilutites (<63μm), calcarenites (63μm–2mm), and calcirudites (>2mm). This is readily applied to particulate rather than crystalline limestones, especially where deposition has been by normal current processes (e.g. the deepwater family of calciturbidites, calcidebrites, calcicontourites, etc).

2. *Folk* (1962, *Mem. Amer. Assoc. Pet. Geol. 1*) (*Fig. 7.1*). This scheme is based on composition and distinguishes three principal components: (a) allochems (particles or grains), (b) matrix (mostly micrite), and (c) cement (mostly sparite). Prefixes for the allochems (bio-, oo-, pel- and intra-) are attached to the root, micrite or sparite, whichever is dominant. Limestones formed *in situ* are called biolithites, and micrites with spar-filled cavities are called dismicrites.

3. *Dunham* (1962, *Mem. Amer. Assoc. Pet. Geol. 1*) (*Fig. 7.2*). The original Dunham scheme classified limestones on the basis of depositional texture into: mudstone (few or no grains), wackestone (coarse grains in a matrix), packstone (grain-support, with matrix), and grainstone (mud absent). Where components are organically bound, the term boundstone is used, and where recrystallization has destroyed the original texture, the rock is referred to as a crystalline carbonate.

Modern field practice is to combine the compositional and textural schemes, so that the most commonly occurring limestone types are oolitic and skeletal grainstones, skeletal and peloidal wackestones, and various reef-limestones or boundstones.

original components not bound together during deposition	mud-supported	less than 10% grains	mudstone
		more than 10% grains	wackestone
	grain-supported		packstone
	lacks mud and is grain-supported		grainstone
original components bound together			boundstone
depositional texture not recognizable			crystalline
original components not organically bound during deposition	>10% grains >2mm	matrix supported	floatstone
		supported by >2mm component	rudstone
original components organically bound during deposition		organisms act as baffles	bafflestone
		organisms encrust and bind	bindstone
		organisms build a rigid framework	framestone

7.2 **Classification of limestones based on depositional texture.**
After Dunham 1962, with modifications by Embry & Klovan, 1971, Bull. Can. Petrol. Geol. 19, 730–781.

Principal characteristics of dolomites

See photographs and figures in the relevant sections of Chapter 3 as well as at the end of this Chapter.

Dolomite formation

Most dolomites have formed by the replacement of limestones: (a) by penecontemporaneous dolomitization soon after deposition, (b) during shallow-burial diagenesis, and (c) during deep-burial diagenesis. Some Precambrian dolomites show no evidence of replacement and may be of primary origin.

Sedimentary features

The sedimentary features of most dolomites, therefore, will closely reflect those of the precursor limestones, with a variable diagenetic overprint.

Penecontemporaneous dolomites are mostly fine-grained and best preserve the original features. They form in semi-arid regions on high intratidal–supratidal flats, preserving desiccation cracks, evaporites and their pseudomorphs, microbial lamination, and fenestrae.

Partial dolomitization of limestones is common during burial diagenesis. Originally aragonitic and high-Mg calcite grains may be dolomitized leaving other parts unaffected. Alternatively, burrows, veins, vugs or particular limestone facies may be the preferred sites for dolomitization. Rhombs of dolomite may occur scattered through the limestone and give a spotty appearance on weathering.

Pervasive dolomitization also occurs during diagenesis, particularly where dolomitizing fluids have followed tectonic structures such as faults, joints, unconformities. This can lead to more complete obliteration of the original structure, texture and composition, yielding a purely crystalline mosaic. Xenotopic mosaics comprise anhedral crystals with irregular, curved boundaries. Idiotopic mosaics comprise euhedral, rhombic crystals with straight edges.

FIELD TECHNIQUES

Although limestones and dolomites can only be studied in a limited way in the field, careful observation will be very rewarding.

- Search the area for the best outcrops.
- Use a hand lens on both fresh and weathered surfaces.
- In addition to the HCl test, remember that Alizarin Red S mixture stains calcite red but leaves dolomite (and quartz) unstained. More information will then be gained by sampling for thin section work and polished slabs.

Occurrence

BOTH limestones and dolomites are common rocks throughout the geological record. Most Precambrian examples are dolomites, Paleozoic rocks are mixed, whereas more recent Phanerozoic (Mesozoic–Cenozoic) carbonates are dominated by limestones. Recent work suggests a further peak of dolomite occurrence around mid-Cretaceous time. At the present day, carbonate sediments are most typical of shallow shelf or platform areas and capping seamounts at mid to low latitudes, as slope deposits surrounding these shallows, and as open-ocean pelagic deposits where the seafloor is above carbonate-compensation depth. They occur more locally in lacustrine and fresh springwater settings. These same environments are well represented in older carbonate rocks. Whereas reef limestones, microbial carbonates, oolites, tufa, and speleothem deposits are each characteristic of very specific environments, many carbonate rocks are not. Fossil and trace fossil evidence, coupled with sedimentary structures and associated facies must be used judiciously in any environmental interpretation.

Field photographs

7.1   Parallel-bedded condensed pelagic limestone succession with 20Ma hiatus (arrow), not recognized in the field but determined from subsequent paleontological study; succession deposited on relative high (bank) between deep-water basins of the former Tethys Ocean.
Jurassic, Umbro-Marche region, central Italy.

7.2   Wavy-bedded limestone-marlstone succession, showing typical rhythmic cyclicity (Milankovitch cycles) of pelagic–hemipelagic sedimentation, outer shelf setting.
Width of view 15m.
Paleogene, near Benidorm, SE Spain.

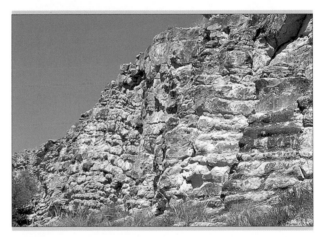

7.3   Interbedded limestone and marl (bedding gently inclined from horizontal), with strong cleavage developed (steeply dipping fabric).
Width of view 1.5m.
Paleogene, Pissouri Basin, S Cyprus.

7.4 Poorly bedded, well-cemented micrite, generally structureless carbonate mud-stones, showing thin darker layers which represent periods with slightly more clay/organic-carbon deposition. Fractured appearance due to weathering. Hammer 45 cm.
Cretaceous, Crimean Peninsula, Ukraine.

7.5 Thin to medium-bedded micrite–biomicrites (mudstone–skeletal wackestone), with prominent vertical joint set. Micrite beds are structureless. Width of view 60cm.
Cretaceous, SE Cephallonia, Greece.

7.6 Nodular-bedded micrites and biomicrites (mudstones and skeletal wackestones) with red dissolution clays; many nodules contain small ammonites as cores. Hammer 30cm.
Jurassic, Rosso Ammonitico, Umbro-Marche, central Italy.

7.7 Bioclastic shell-rich muddy limestone (marlstone or wackesetone), fragmented shell debris, indistinctly bedded. Width of view 30cm.
Pliocene, W Cephallonia, Greece.

7.8 Microbial paper-micrites (mudstone to bindstone), with highly fissile lamination, probably from lagoonal depositional setting. Note minor fault with little displacement.
Width of view 30cm.
Late Miocene, Pissouri Basin, S Cyprus.

7.9 Oolitic limestone (oosparite or oolitic grainstone) with indistinct parallel lamination. The individual sand-size ooids typically show up most clearly on partially weathered surfaces, as is the case here. Key 6cm.
Carboniferous, West Yunnan, China.

7.10 Sandy oolitic limestone (oosparite or impure oolitic grainstone) with bioclastic debris and terrigenous fraction (note grey, glassy quartz grains); note also the bioturbated fabric. Knife 10cm.
Jurassic, Osmington Mills, S England.

7.11 Muddy oncolitic limestone (oncolitic packstone); shallow-marine to tidal environment. Note concentric microbial lamination in individual oncolites, mostly spherical to oval in shape. Width of view 20cm.
Jurassic, Osmington Mills, S England.

7.12 Intraclast and oncolitic sparitic limestone (or packstone); calcirudite formed by off-bank resedimentation from shallower water. Coin 2.5cm. *Jurassic, Umbro-Marche region, central Italy.*

7.13 Biosparite (skeletal packstone) with mixed biota including bivalves, gastropods, coral fragments, and microbial material; talus breccia (or calcirudite) from adjacent bioherm. Coin 2.5cm. *Neogene, Roldan reef complex, Carboneras, SE Spain.*

7.13 Detail of fallen *Porites* coral fragment in talus-slope calcirudite (rudstone). Width of view 25cm. *Neogene, Cabo de Gata, SE Spain.*

7.15 Biolithic (framestone) branching corals *in situ*, together with microbial growth and bioclastic debris. Lens cap 6cm. *Pleistocene, Island of Guam, W Pacific.*

7.16 Microbial laminated sediment and low-relief mounds (stromatolites) over pink-coloured microbial-bound, homogeneous material (thrombolites); making up an *in situ* microbial biolithite (bindstone to framestone). Hammer 25cm. *Silurian, Murrumbidgee, ACT, Australia.*

7.17 Microbial biolithite (bindstone) forming small reef knoll (part). Note that much of the carbonate is in fact strontianite ($SrCO_3$) – dense and heavy to hold – and calcio-strontianite, as a result of diagenesis over a thick evaporite unit. Lens cap 6cm. *Latest Miocene, near Khirokitia, S Cyprus.*

7.18 Moderate-relief stromatolite mounds (near base) overlain by coarse biosparite with microbial binding (framestone passes upwards into bindstone). Hammer 25cm. *Silurian, Murrumbidgee, ACT, Australia.*

7.19 Isolated stromatoporoid mound (member of the *Porifera* phylum) within thin-bedded micrite (carbonate mudstone to skeletal wackestone). Hammer 25cm. Photo by Paul Potter. *Ordovician, Fayette County, Kentucky, USA.*

7.20 Part of Waulsortian carbonate mud mound in the Fort Payne Limestone Formation. Such carbonate buildups or bioherms are mostly Paleozoic and occur within deeper-water muds and limestones. Typically they are composed of structureless micrite, and biomicrite with scattered skeletal debris (mudstone and skeletal mudstone).
Photo by Paul Potter.
Lower Carboniferous (Mississippian), Kentucky, USA.

7.21 Jumbled blocks of bioherm debris in reef talus-slope calcirudite limestone (floatstone to mudstone), over well-bedded calcilutites (micrites or carbonate mudstones). Hammer 45cm.
Miocene, Pissouri Basin, S Cyprus.

7.22 Biomicrite (skeletal wackestone) showing partial silicification (irregular, glassy, grey patches). Deep-water basin succession. Coin 2.5cm.
Jurassic, Umbro-Marche region, central Italy.

7.19 Calcirudite turbidite (sparite or grainstone) erosive into pelagic calcilutite (sparry micrite) with *Zoophycos* trace (Z). Bedding picked out by thin dissolution seams. Deep-water slope to basin succession. Hammer 45cm.
Cretaceous, Monte Conero, central Italy.

7.21 Parallel-bedded calcilutites (micrites or mudstones); distal turbidites (T, arrows) and pelagites (P) in deep-water slope to basin setting. The distinction between turbidite and pelagite is often very difficult to make in the field, as is the case here. Lens cap 6cm.
Paleogene, Lefkara, S Cyprus.

7.25 Calcilutite and calcarenite showing irregular lenticular bedding and subtle grain-size oscillation; mainly biomicrite (packstone to grainstone) with minor terrigenous fraction; interpreted as contourites from detailed analyses and the overall context (not possible from visual identification alone).
Width of view 25cm.
Paleogene, Lefkara, S Cyprus.

7.26 Bioclastic calcarenite (skeletal packstone to grainstone) with alternating beds of parallel-laminated and medium to large-scale cross-stratified sets (sets up to 0.8m); shallow-marine, probable tidal setting.
Width of view 7m.
Cretaceous, Bonafaccio, Corsica, France.

7.27 Inorganic limestone cold-water tufa deposit (mainly sparite) within and overlying soil horizon. Hammer 30cm.
Recent, Mascarat, SE Spain.

7.28 Detail of tufa – a spongy-textured, coarsely crystalline limestone, typically formed around springs, coating rootlets and other plant material, and with some organic (microbial) influence on precipitation. Note subparallel growth bands and many elongate cavities.
Width of view 8cm.
Recent, Jura Mountains, France.

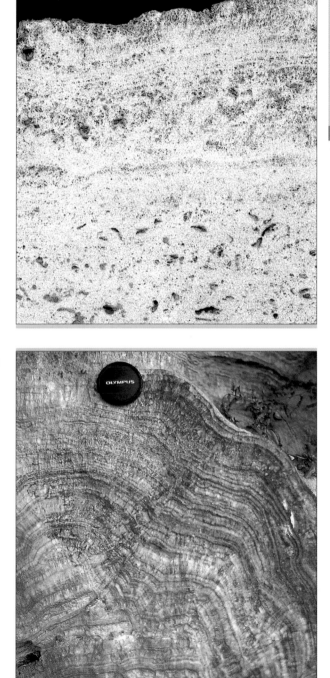

7.29 Detail of travertine – a dense, compact, finely crystalline limestone, typically formed as flowstone and dripstone deposits in caves or cavities, and as deposits around springs (as in this view). Note concentric lamination and radial pattern of crystal formation. May occur as purely an inorganic precipitate, although also with microbial influence on precipitation.
Lens cap 6cm.
Recent, West Yunnan, SW China.

7.30 Detail of travertine – inorganic limestone karst precipitate (sparite) within former cave or cavity system (speleothem). Coin 2.5cm.
Age uncertain, near Carboneras, SE Spain.

7.31 Massive dolomite – a very hard rock that does not effervesce readily with dilute HCl. Typically makes a sharp ring when struck with a hammer. Width of view 25cm.
Triassic, Dolomites, N Italy.

7.32 Interbedded dolomite (partly brecciated) and limestone (micritic). Dolomite breccia occurs as intraclast bed with minor erosive scour; intraclasts formed by reworking of dried-out lagoonal crust. Other evidence of original evaporative lagoon include salt pseudomorphs and limited euryhaline fauna. Width of view 80cm. Photo by Ian West.
Cretaceous, Dorset, S England.

7.33 Selective dolomitization of biomicrite, indicated by the brown-coloured patches. This colour staining is from oxidation of the iron substitution in dolomite lattice. Lens cap 6cm. *Carboniferous, SW Yunnan, China.*

7.34 Dolomite, known locally as Magnesian Limestone. Brownish colour from oxidation of iron, and terrigenous sandy appearance due to individual crystals of dolomite being precipitated directly from hyperconcentrated seawater brines and then washed gently by nearshore currents. Width of view 6cm. *Triassic, Nottingham, UK.*

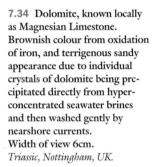

7.35 Dolomitized reef talus biosparitic limestone, now very hard, crystalline, pinkish coloured and with many fractures and cavities. The pinkish colour is due to oxidation of iron replacing magnesium in the dolomite lattice. The cavities may form, in part, because the dolomite crystal size is 12% smaller than the calcite it replaces, but may also be due to other factors. Width of view 15cm. *Triassic block in Mamonia Melange, Aphrodites Bay, S Cyprus*

CHERTS
AND SILICEOUS SEDIMENTS

Cyclic bands of dark flint nodules in Cretaceous Chalk, Dorset coast, UK.

Definition and range of types

CHERT is the general term for fine-grained siliceous sedimentary rock of biogenic, biochemical or chemogenic origin. It is made up of fine-grained silica, with only small quantities of impurities. By contrast, modern siliceous sediments may have a high proportion of impurities, either biogenic carbonate or siliciclastic muds. Extremely pure siliceous oozes are a relatively rare deep-sea facies, so that the process of producing pure chert generally requires diagenetic purification.

Cherts are generally divided into bedded and nodular types (Table 8.1). The former mostly develop from primary accumulations, whereas the latter have a diagenetic origin

that does, nevertheless, reflect primary deposition of siliceous sediment. Both are very common throughout the geological record.

Particular types of chert have been given specific names: flint is a very fine-grained, mostly black nodular chert common in Cretaceous chalks; jasper is a red chert, the colouration due to finely disseminated hematite; black chert is a dark-coloured, organic-rich variety that may occur in both bedded and nodular form. Impure cherts, where there is a significant or even dominant admixture of clays or carbonate, are referred to as siliceous mudstones and siliceous limestones respectively.

Present-day siliceous-rich sediments cover large areas of ocean floor, particularly beneath areas of high organic productivity, and also occur in some lakes. Relatively pure siliceous oozes are known as radiolarian or diatomaceous oozes, depending on the dominant microfossil component. Less pure siliceous sediments are informally known as sarls (siliceous muds) and smarls (siliceous calcareous muds). Porcellanite is a hard siliceous sediment, in which the original biogenic silica (opal-A) has transformed to the more ordered opal CT, but not completely to quartz. Diatomites, radiolarites and spiculites are also partly lithified sediments still rich in opal CT.

Principal sedimentary characteristics

See photographs and figures in the relevant sections of Chapter 3 as well as at the end of this chapter.

CHERTS are generally very hard, dense rocks, which splinter with a conchoidal fracture when struck. They are subject to intense diagenetic recrystallization and remobilization that typically partially or completely obscures primary depositional features. Of the two main types, bedded cherts retain more of their original characteristics, including lamination, cross-lamination, grading, etc. Colour depends on impurities, and ranges from white/pale grey (pure), through browns and reds (traces of iron), greens (traces of iron from chlorite or smectite clays in volcaniclastics), to darker greys and black (clays and organic carbon).

Bedding
Bedded cherts are generally 1 – 10cm thick, but may be >30cm thick, with slightly irregular to wavy margins, commonly appearing within mudrock, volcaniclastic or limestone successions. Nodular cherts occur as ovoid to irregular nodules, very variable in size (long diameter typically 1 – 30cm), commonly concentrated along bedding planes, and in some places coalescing to form semi-continuous irregular layers.

Structures
Bedded cherts are typically massive (either primary or diagenetic) or finely laminated (seasonal or current influence); cross-lamination, graded bedding and micro-scour structures are indicative of turbiditic cherts. Slump folding and synsedimentary brecciation are also evident in some cases.

Primary structures are mostly absent or poorly preserved in nodular cherts. In some cases the nodules form preferentially around a burrow network (e.g. *Thalassinoides* burrows), so that their overall form shows the trace fossil structure.

Texture
Many bedded cherts are dominated by an equant mosaic of microquartz crystals, together with some chalcedonic quartz, that effectively obscure original grain size characteristics; some do reveal original sand–silt–clay sized particles, including grading.

In nodular cherts, the original grain size texture is usually destroyed as dispersed biogenic silica dissolves and reprecipitates as opal-CT at nodule growth points. At a later stage this matures to quartz.

Composition
In bedded cherts of biogenic derivation the original components are mainly siliceous microfossils (radiolarians, diatoms and sponge spicules), plus siliciclastic and calcareous impurities and, in some cases, relatively high traces of organic carbon and phosphates. Radiolarians and sponges occur in Cambrian to Recent sediments, whereas diatoms first appeared in the Triassic, remaining rare until the Cretaceous. These siliceous components are generally recrystallized to quartz, although careful examination may reveal remnant sand-sized spheres, discs and spines.

Nodular cherts are dominated by pure microquartz, macroquartz and chalcedonic quartz formed both as replacement and pore filling cement within the carbonate (or other) host sediment.

Table 8.1 Principal types of chert

Major types	Subtypes	Nature and origin
Bedded chert	Radiolarian chert, diatomaceous chert, spicule-rich chert, black chert, jasper	Mainly marine (some lacustrine) in origin; comprise part recrystallized quartz and part biogenic remains, plus minor impurities.
Nodular chert	Flint	Nodular chert found in Chalk.
	Silcrete	Nodular or encrusting chert formed in certain soils or as surface coating.
Partly lithified siliceous sediments	Radiolarite	Rich in radiolarians.
	Diatomite	Rich in diatoms.
	Spiculite	Rich in sponge spicules.
	Porcellanite	Hard siliceous sediment (general term).
Unlithified siliceous sediment	Siliceous ooze	Soft unconsolidated siliceous sediment (general term).
Hybrid siliceous sediments	Sarl	Siliceous mud (unconsolidated).
	Smarl	Siliceous calcareous mud (unconsolidated).
	Siliceous limestone	Siliceous carbonate sediment.
	Siliceous tuff	Siliceous volcaniclastic sediment.

Occurrence

CHERTS occur in minor amounts throughout the geological record. In bedded cherts of Precambrian successions and in association with volcaniclastic sediments (especially those of acid or intermediate derivation), the silica may, at least in part, have a direct volcanic origin. Other bedded cherts are most likely of primary biogenic origin, produced by the pelagic rain of siliceous planktonic organisms common below regions of high surface productivity. In some cases, shelf and upper slope siliceous sediments have been resedimented downslope yielding chert turbidites. Nodular cherts, on the other hand, are mostly of diagenetic origin. They typically occur within limestone hosts, and less commonly in mudrocks. The silica was originally biogenic and more widely dispersed through the pelagic or hemipelagic sediments. Primary biogenic lacustrine cherts and silcretes within soil horizons occur more rarely.

FIELD TECHNIQUES

Cherts and other hard siliceous sediments have a distinctive ring when struck sharply with a hammer. They also show characteristic conchoidal fracture when broken. However, be very wary of sharp splinters that can fly off when struck, and are especially dangerous to unprotected eyes.

8.1 Core sections split lengthways showing parts of subrecent siliceous debrite unit, originally deposited as pelagic biogenic facies under the upwelling zone on Walvis Ridge. Sub-rounded and soft-sediment deformed clasts of organic-rich, siliceous facies (sarls, dark), and organic-poor, calcareous facies (marls–oozes, pale).
Width of core 8cm.
Plio–Pleistocene, Walvis Ridge, SE Atlantic Ocean.

8.2 Irregular, nodular band of dark flint within chalk. The bands are bedding parallel, and represent periods of time when there was an increased pelagic rain of siliceous tests to the sea floor. Dissolution, remobilization, and precipitation of the silica as irregular nodules occurred during burial diagenesis. Width of view 35cm.
Late Cretaceous, Isle of Wight, S England.

8.3 Detail of flint nodule within chalk. Note the white chalk (micrite) coating and dark grey interior of nodules. *See also* caption for **8.2**. Width of view 20 cm. *Late Cretaceous, Isle of Wight, S England.*

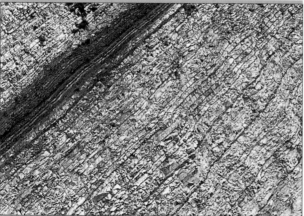

8.4 Black organic-carbon-rich chert beds in limestone succession. The chert beds represent periods of time with dominant accumulation of siliceous biogenic material at the seafloor. Recrystallization during burial diagenesis has been largely *in situ*. The black colour in this case is due to preservation of organic carbon. Width of view 6m. *Mid-Cretaceous, Umbro-Marche region, central Italy.*

8.5 Thin chert bands/beds (pale) in dark sparitic limestone. Note dissolution seams between beds. Penknife 9cm. *Paleogene, central Crete.*

8.6 Chert concretion (C) in dark sparitic limestone (as 8.5); note incipient formation of bedding-parallel stylolites (S) in limestone. Penknife 9cm.
Paleogene, central Crete.

8.7 Large concretions of dark grey chert in limestone. Hammer 50cm.
Photo by Paul Potter.
Lower Carboniferous (Mississippian), Gilbertsville, Trigg County, Kentucky, USA.

8.8 Calcilutite showing partial silicification in lenses (pale grey) and nodules (darker grey). Lens cap 6cm.
Jurassic, near Scaglia Bani, Sicily, Italy.

8.9 Organic-carbon-rich radiolarites and siliceous micrites; well-laminated upper section is mainly pelagic; more irregular beds are debrite–turbidite (D,T) intercalations. *Miocene, Monterey Formation, California, USA.*

8.10 Thick-bedded radiolarite turbidite–debrite (arrow), with coarse clast-rich base and large floating clasts (F) and raft (R) about 1.5m up from base. *Miocene, Monterey Formation, California, USA.*

8.11 Interbedded micrite and chert (darker layers, yellowish in parts). Many of the chert beds are of turbidite origin (lamination and grading visible on close inspection), deposited within a mainly pelagic carbonate slope-basin setting. The yellowish colour derives from terrigenous impurities. Note cross-cutting faults with minor displacement. Width of view 1.2m.
Paleogene, Lefkara, S Cyprus.

8.12 Chert beds (pale yellowish) in micrite; cherts are of turbidite origin deposited within a pelagic carbonate slope-basin setting. Beds steeply inclined; exposure covered with much loose rubble. See also caption for 8.11. Width of view 3.5m.
Paleogene, Lefkara, S Cyprus.

PHOSPHORITES

Favoured perch for cormorants leads to guano accumulation, central Chile. *Photo by Claire Ashford.*

Definition and range of types

SEDIMENTS and sedimentary rocks that are composed largely of phosphatic material are collectively known as phosphorites. Although phosphorus is an essential component of all living matter, and the dominant constituent of vertebrate bones, phosphorites are relatively rare in the sedimentary record.

Phosphorites are generally divided into three main groups (Table 9.1): bedded and nodular phosphorites, usually of marine origin and formed in association with high organic productivity; bioclastic lag phosphorites (bone beds); and guano (bird or bat excrement) and its associated facies. Guano is not known from the ancient geological record and is not discussed further here.

Principal sedimentary characteristics

See photographs and figures in the relevant sections of Chapter 3 as well as at the end of this chapter.

PHOSPHORITES are relatively rare and not easily recognized in the field. They mostly occur as dark organic-rich sediment, slightly enriched in phosphate and as distinct phosphatic nodules and crusts, with a dull black to yellow-brown coating. Bioclastic lag deposits are more restricted in occurrence. They are generally dark brownish-black, shiny, hard, and relatively resistant to weathering, but may become friable on breaking.

Bedding

The bedded/nodular phosphorites occur either as well-bedded layers, generally as thin beds; or as spherical, ovoid, tubular, and irregular nodules (typically 1–10cm diameter), commonly in distinct layers. Associated facies include phosphatic mudrock, dolomite, and limestone. Bioclastic lag phosphorites are invariably well-bedded, and of variable thickness (generally <30cm).

Structure

Beds and nodules may display homogeneous or laminated structure, or a more chaotic conglomeratic form. Lag deposits are typically more jumbled or chaotic, but with bed-parallel and current-parallel alignment of clasts evident in many cases.

Texture

Grain size varies from fine-grained mudrock to sand-sized oolitic, pisolitic, and pelletal texture. Concentrations of phosphatic nodules may appear conglomeratic. Lag deposits are generally very poorly sorted, of variable grain size, from fine mud matrix to elongate fecal pellets (up to 2cm), and larger phosphatic pebbles.

Table 9.1 Principal types of phosphorites

Major types	Subtypes	Nature and origin
Bedded and nodular phosphorites	Nodules and pavements, bioclastic phosphorites, pelletal and oolitic phosphorites, phosphate-rich muds	Formed in association with organic-carbon rich sediments under upwelling systems and enhanced organic productivity; open marine outer shelf/upper slope setting; range of facies depending on local conditions and associated sediments.
Bioclastic lag phosphorites	Phosphatic bone beds	Skeletal fragments (vertebrate bones, fish scales) and coprolites concentrated by currents/waves. Associated with slow deposition or hiatus.
	Phosphatic pebble beds	Reworked bone beds and (diagenetic) phosphate nodules.
Guano	Bird guano	Localized, thick accumulation of bird excrement.
	Bat guano	Localized, thick accumulation of bat excrement.
		(Both can be associated with phosphatization of underlying rocks.)

Composition

Most sedimentary phosphorites are carbonate hydroxyl fluorapatites $[Ca_{10}(PO_4,CO_3)6F_{2-3}]$; in some cases with a distinct mineralogy (francolite, F-rich; dahllite, F-poor) but more often as the cryptocrystalline collophane. The P_2O_5 content ranges from 5–35%.

In lag deposits, there are a wide variety of phosphatic clasts, phosphatized bioclastics and non-phosphatic material; faecal pellets and fish scales are also common.

Occurrence

MARINE PHOSPHORITES are well known from present-day and sub-recent occurrences along the outer shelf/upper slope regions of continental margins that have enhanced organic productivity. They mostly form in low-oxygen rather than fully anoxic conditions, occurring as phosphatic pellets, nodules, crusts, and thin beds; Evidence of current reworking is common. Ancient examples of this type are known from the Precambrian onwards, with apparent peaks in occurrence during the late Precambrian–Cambrian, the Permian, the late Cretaceous–early Tertiary, and the Miocene–Pliocene periods. These were periods of high sea level and widespread transgression leading to shallow, fertile shelf seas. Phosphorites are typical of condensed sequences, extremely low rates of sedimentation, and the development of hardgrounds. Bioclastic lag deposits are more localized in occurrence, but significant as evidence of slow accumulation or hiatuses.

> FIELD TECHNIQUES
> Phosphorites are rare and not easily recognized in the field. If you suspect their occurrence, take samples back to the lab for thin-section and chemical analyses.

Field photographs

9.1 Phosphorite concretion (dark centre with white powdery surround) in mudrock. Lens cap 6cm. *Paleogene, London Clay, S England.*

9.2 Phosphorite concretions (brownish-coloured) in dark-grey, organic-rich mudrock. Hammer 30cm.
Cretaceous, Speeton Clay, S England.

9.3 Detail of phosphorite concretions collected from various locations. Mostly brownish coloured, some with darker grey interior.
Width of view 15cm.
Jurassic, Oxford Clay, S England.

9.4 Phosphorite concretions around crab carapaces – the left one found *in situ*, the right one found reworked in glacial outwash deposit. Note the dull to shiny black colour typical of phosphatized remains.
Width of view 10cm.
Paleogene, London Clay, S England.

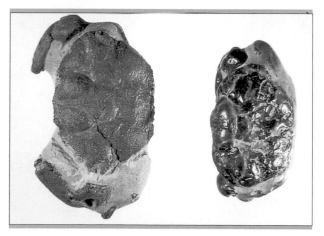

9.5 Detail of phosphorite nodule, in which the phosphorite (francolite) is encasing and replacing limestone in glauconitic marl. The black-green grains are glauconite, which commonly co-occurs with phosphorite.
Width of view 12cm
Cretaceous, Isle of Wight, S England.

9.6 Limestone hardground, phosphatized and overlain by glauconitic marl. Knife 9cm.
Cretaceous, Isle of Wight, S England.

9.7 Bedding-plane view of phosphatic horizon within a condensed limestone unit, known locally as the Ludlow Bone Bed; the numerous bioclastic fragments are mainly phosphatized denticles and spines from a primitive agnathid vertebrate. Width of view 10cm.
Silurian, Shropshire, England.

9.8 Phosphorite concretions (brownish-yellow colour) in glauconitic sandstone. Width of view 10cm.
Age uncertain, Gambia.

9.9 Phosphorite and carbonate nodules in marlstone.
Age uncertain, Gambia.

COAL AND OIL

Outcrops of Middle Pennsylvanian coals, Eastern Kentucky. *Photo by Jim Hower.*

Definition and range of types

COAL AND OIL (including natural gas) are the two principal fossil fuels found in sedimentary rocks and, currently, the world's dominant source of commercial energy. They have close parallels in their origin but significant differences in occurrence. Whereas coal is a common sedimentary rock, oil is a liquid fuel that is hosted or reservoired in sediments. They are also closely linked in that coal is an important source rock for gas and, in some cases, for oil too. Coal is the main focus of this chapter, but the origin and occurrence of oil is also described briefly.

Coals are brown to black, soft to hard, low-density, organic-carbon-rich sedimentary rocks. By definition they should contain <33% inorganic material (impurities) but generally they have <10% clay, silt, and sand, as well as a host of minerals present in minor and trace amounts. They are a very common (and economically important) sediment type that is easily recognized in the field.

Most coals are humic coals, formed from the *in situ* accumulation of plant material. Sapropelic (or drift) coals, by contrast, have formed from the dispersal and subsequent accumulation of broken-up plant debris, spores, pollen, and/or algae. Coal rank refers to the degree of metamorphism (or coalification) of the organic matter. Humic coals can be divided on the basis of their rank into a number of categories.

These are:

- *peat* (lowest rank)
- *lignite* (soft brown coal)
- *sub-bituminous coal* (hard brown coal)
- *bituminous coal* (hard coal)
- *semi-anthracite*
- *anthracite* (highest rank)

Each of these categories (or classes) can be further subdivided into coal groups (Table 10.1). The various properties used to measure rank all require laboratory analysis, although some indication can be obtained from field examination.

Sapropelic coals, though also showing an evolution in rank as a result of burial metamorphism, are generally classified differently. Cannel coal is the typical massive, finer-textured, mixed-composition type, whereas boghead coal is largely derived from algae.

Oil (or crude oil) is the liquid form of petroleum, whereas gas (or natural gas) is the gaseous form. Crude oil is formed from the thermal alteration of certain types of organic matter during the progressive burial of sediment containing significant amounts of organic matter. These organic-rich source rocks are mostly fine-grained, dark-coloured mudrocks, known as black shales and oil shales (see pages 152–154) and, more rarely, fine-grained limestones and cherts. The main phase of oil generation occurs at temperatures of 70–120°C, typically at burial depths of 1–3km. This is known as the oil window. Gas begins to form within the oil window, but then continues to form at slightly higher temperatures and greater burial depths. All gas and most lighter oils escape easily at the surface and are not, therefore, seen in sedimentary rocks exposed at outcrop. Heavier oils, however, do remain behind, giving a dark, brownish colour to the sediment, with leakage of black bitumens from joints and surface cracks. Where these heavy oils are particularly abundant in arenaceous sediments, they are known as tar sands (see pages 140–142).

Principal sedimentary characteristics

See photographs and figures in the relevant sections of Chapter 3 as well as at the end of this chapter.

COALS are well known sedimentary rocks in formations of all ages since late Devonian times, shortly after land plants first evolved and proliferated. They are easily recognised in the field, on the basis of their black colour, light weight, and associated rootlet beds, and yield important environmental information. However, many of their specific features are more readily defined through routine sampling and laboratory analysis. Much sedimentological work on coals is carried out on cores recovered from shallow and deeper boreholes.

Bedding

Coal seams (beds) tend to be thinner (<3m) but more laterally extensive in deltaic settings, while thicker and more restricted in fluvial sequences. Some of the thickest coal seams (up to several hundred metres) are found in fluvial and alluvial-fan systems. Even very thin seams (<5cm) may have an extensive rootlet system. Thicker, well-established coal seams and rootlet systems may have leached the underlying sediment (sand or mud) giving a distinctive, generally pale-coloured, seat-earth layer.

Structures

No dynamic sedimentary structures are apparent, but millimetre to centimetre-thick layering of different microlithotypes is common. These are known as bands. A somewhat subjective and time-consuming but very useful method for the macroscopic description of coals involves the visually logging of bands in coal seams, both in the field and in cores. Each seam may develop a distinctive banding profile, which can facilitate seam identification and correlation across the coal basin. The *Australian Brightness Profile* for band measurement, modified from the Mary Stope's classification of lithotypes

Table 10.1 Principal types of coal

Major types (humic coals)		Rank parameters (approximate average values or range)					
CLASS	GROUP	%C	%V	VR	%W	Cal	HCgen
Peat		<50	>50	–	>75	<12	
Lignite	B	50–55	52	0.3–0.36	55–75	12–14.7	
	A	55–60	47	0.36–0.42	35–55	14.7–16.3	early gas
Sub-bituminous	C	60–65	42	0.42–0.45	30	19.3–22.1	
	B	65–70	39	0.45–0.48	25	22.1–24.4	
	A	70–75	37	0.48–0.50	15	24.4–26.8	
Bituminous	High volatile	75–80	33	0.50–1.12	<10	26.8–34.0	oil and gas
	Med. volatile	80–82.5	26	1.12–1.51			
	Low volatile	82.5–85	18	1.51–1.92			wet gas
Anthracite	Semianthracite	87	11	1.92–2.50			
	Anthracite	90	5	>2.50			dry gas
	Meta-anthracite	92.5	<2				

Other types (sapropelic coals) including cannel and boghead groups – variable properties

%C	= weight percent carbon
%V	= weight percent volatile matter (dry mineral-matter free)
VR	= vitrinite reflectance
%W	= weight percent water
Cal	= calorific value MJ/kg (water and mineral-matter free)
HCgen	= equivalent onset stage of hydrocarbon generation

Classification scheme after Marlies Teichmuller (1975)

(see *Composition* below), characterizes bands as follows:

- *C1* Bright coal (bright bands >90%).
- *C2* Bright banded coal (bright bands 90–60%).
- *C3* Banded coal (bright bands 60–40%).
- *C4* Banded dull coal (bright bands 40–10%).
- *C5* Dull coal (bright bands <10%).
- *C6* Dull with satin sheen, friable, with up to 10% other coal lithotypes.

Vertical joints (known as cleats in coal seams) occur in many coals, often coated with different minerals. There may also be evidence of synsedimentary faulting and/or bed-parallel slippage surfaces in others. Partings of other rock types (e.g. carbonaceous mudstone, sandstone, volcanic ash) may be present within coal seams, and further aid in identification and correlation. Rootlet horizons are common below coal seams and will even occur in the absence of a preserved coal layer. Their presence in a sedimentary succession may indicate the occurrence of laterally adjacent coals.

Texture

Coal is a compact fine-grained rock, of equivalent grain size to mudrocks. For the most part, therefore, individual grains are not visible in the field, even with a hand lens. Polished thin sections in the laboratory show that some of the component macerals (see below) can be silt or sand sized. In peat, the coarser fragments of undecomposed plant material may be visible.

Composition

The fundamental (microscopic) components of coals are known as macerals, which are derived from different parts of the original plant material (Table 10.2). These can vary significantly for different types and ages of coal. Maceral assemblages characterize specific microlithotypes, which when visible in hand-specimen are known as coal lithotypes. The most common lithotypes are:

- *Vitrain:* bright bands, glassy, brittle, conchoidal fracture.
- *Clarain:* both bright and dull bands, finely laminated, silky, with a smooth fracture.
- *Durain:* dull bands, hard, without lustre.
- *Fusain:* charcoal-like, soft, powdery, soils the fingers.

With increasing rank, the macerals tend to lose their specific character and the coal becomes more homogeneous and progressively harder. Macroscopic plant material (fossils) can only be clearly recognized in the lower-rank coals – peat and lignite.

A variety of inorganic components may also be present in coal. These include detrital quartz, clay minerals, heavy minerals, sulphates and phosphates, as well as diagenetic nodules of pyrite (very common), marcasite, siderite, ankerite, dolomite and calcite.

Other field observations

Certain other features can be very important to coal studies. The type and nature of underlying and overlying strata (sandstone, mudstone, etc.), as well as the ratio of coal to non-coal sediments, are significant to mining engineers. So too is the coal geometry – rapid thinning/thickening, wedging of seams, and washouts. The presence of any igneous rocks, intrusive or extrusive, may have caused the coal to be heat affected, over-rapidly matured, and hence to deteriorate in quality. Any features that might aid correlation, such as the presence of partings, ash layers, diagenetic nodules, and bands, are important to note. Synsedimentary faulting, as well as observations on cleat directions and other structural features, will help with overall interpretation of the coal basin.

Table 10.2 Principal components of coal and their origin

Principal macerals	Primary origin
Vitrinite group	Derived from cell-wall material (woody tissue) of plants composed of polymers, cellulose, and lignin.
Liptinite group	Derived from waxy and resinous parts of plants, including pores, cuticles and resins, and also algae. Sensitive to advanced coalification and usually disappear from medium to low-volatile rank bituminous coals.
Inertinite group	Derived from plant material that has been strongly altered and degraded during the peat stage of coalification by oxidation. Includes fossil charcoal (known as fusinite).

Occurrence

Coals

Coal facies are mostly associated with both paralic (deltaic and coastal) environments, and with limnic (fluvial and lacustrine) successions, and do not occur within fully marine systems. They form from the accumulation of mainly terrestrial plant material and its preservation under anoxic or low-oxygen conditions. This occurs most readily in humid climates. The most abundant coals are of late Carboniferous and Permian age in the northern hemisphere (especially North America and Europe), mostly occurring within paralic successions. Permian age coals in limnic successions are typical of many southern hemisphere occurrences formed on the former continent of Gondwanaland. Other peaks of coal formation occur worldwide, and especially in China, in the Jurassic, Cretaceous, and mid-Tertiary periods. Tertiary coals are typically lower rank lignitic or brown coals. Rare, early Paleozoic and Precambrian coals are exclusively of algal derivation.

Carboniferous northern-hemisphere coals are typically rich in vitrinite group macerals (60–90%), together with 5–15% liptinites and 5–40% inertinite. Permian Gondwanaland coals, and some west Canadian coals, are vitrinite-poor and correspondingly richer in inertinite. Tertiary coals are typically lower-rank lignitic or brown coals, except where deeply buried or much influenced by igneous/tectonic heating. Cannel and boghead coals are dominated by liptinite macerals. Rare, early Paleozoic and Precambrian coals can also be found, and are exclusively of algal derivation.

Oil and gas

Most oil forms from marine organic matter (much of planktonic origin); some forms from lacustrine algal organic matter. Most gas forms from terrestrial organic matter, such as the higher plant material in coals. Both require the accumulation and preservation of organic matter under anoxic or low-oxygen

> **FIELD TECHNIQUES**
> Coal seams (or beds) are easily identified in the field but difficult to examine in detail without laboratory analysis. Carbonaceous (coal) fragments are common in many other sediments and, where present in abundance, they can indicate proximity to the coal source. However, I have encountered common carbonaceous fragments in Recent sandstone turbidites of the Indian Ocean, at 5000m water depth, some 2500km from their source on the Ganges Delta – so beware!

conditions. The dominant source rocks around the world are Jurassic, Cretaceous, and mid-Tertiary in age.

Once formed, over 90% of the petroleum generated in sediments escapes at the surface through natural seepage. Much of the remainder migrates into hydrocarbon traps comprising porous reservoir rocks of almost any sediment facies, of which sandstone and limestone reservoirs are the most common. Reservoirs of all ages are known and exploited, but there is an overwhelming dominance of Tertiary and Mesozoic reservoirs. Some oil never escapes from its source rock, due to lack of permeability, and therefore remains tightly bound within the complex kerogen structure to form an oil shale.

As oil migrates through a sedimentary basin, some will migrate into traps while much continues through permeable conduits and eventually escapes at the surface. This process tends to lead to hydrocarbon fractionation, such that the gas and lighter oils are trapped in the deeper, more central parts of basins, while the heavier oils are located around the margins. The very heavy oils associated with tar sands are generally found at basin margins, and have been further degraded by bacterial action and water flushing.

Field photographs

10.1 Very thick coal seam within dominantly sandstone succession; thinner coal seams up section. Approximately 300 dip of strata from upper right to lower left.
Carboniferous, Cardinal River Opencast Site, Rocky Mountains, Canada.

10.2 Medium-thick coal seams within sandstone-mudstone synclinal succession.
Carboniferous, Westfield Opencast Site, Fife, Scotland.

10.3 Very thin, discontinuous coal horizon (centre of view), underlain by fine carbonized rootlet traces; deltaic–coastal succession. Hammer 25cm.
Paleogene, near Fribourg, Switzerland.

◄ **10.4** Medium-thick coal seam draping low-relief scour surface; associated with dark carbonaceous shales and thin-bedded sandstones.
Photo courtesy of Stephen Greb, Kentucky Geological Survey.
Carboniferous, US23, Kentucky, USA.

▶ **10.5** Thin coal seam and underlying rootlets within deltaic sandstone succession. Width of view 30cm.
Jurassic, near Whitby, NE England.

▶ **10.6** Thin to very thin coal seams (C) with rootlets, and rootlet traces within original seat earth beds above which the 'coal' has not been preserved. Brown-orange staining is due to weathering of iron-rich minerals to iron oxides/hydroxides.
Jurassic, near Whitby, NE England.

◄ **10.7** Medium-thin coal seam capping small-scale scour; associated with dark carbonaceous shales and thin-bedded sandstones.
Photo courtesy of Stephen Greb, Kentucky Geological Survey.
Carboniferous, US23, Kentucky, USA.

10.8 Detail of anthracite coal, *Carboniferous, S Wales.*

10.9 Detail of bituminous coal, *Carboniferous, S Scotland.*

10.10 Detail of lignitic coal, *Tertiary, SW China.*

10.11 Oil shale, composed of highly organic-rich, biogenic siliceous mudstone (diatomite). This is now saturated with heavy oils and oozes black tar over the steeply dipping rocks and across the beach. The most oil-rich shales in this succession are, in fact, diatomite and diatomaceous mudstone.
Miocene, S California, USA.

10.12 Detail of oil shale (as in 10.11). Finely laminated organic-rich diatomaceous mudstone showing natural oil seepage. Lens cap 6cm.
Miocene, S California, USA.

10.13 Tar sand, composed of heavy oil-saturated shallow marine sandstone, hence general dark colour. Note natural tar oil seeping from thin sub-vertical joints. The brownish-red banding across lower field of view is caused by late-stage diagenetic flow.
Width of view 40cm.
Jurassic, Osmington Mills, S England.

EVAPORITES

Secondary Triassic gypsum showing deformation in thrust plain, Pyrenees, Spain. *Photo by Ian West.*

Definition and range of types

EVAPORITES are chemogenic sediments that have been precipitated from water following the concentration of dissolved salts by evaporation. This takes place from both marine and non-marine (lacustrine, lagoonal) waters.

Although there are a very large number of different chemical salts dissolved in seawater, their relative abundances and solubilities allow only very few common evaporite minerals to precipitate naturally. These include halite, gypsum, anhydrite and, to a lesser extent, dolomite and potassium salts (or bitterns). Prolonged evaporation leads to significant accumulation of evaporite minerals, and the resulting sedimentary deposit is generally named after the dominant mineral. In thick successions there is commonly a cyclic repetition of evaporites from less to more soluble – that is, dolomite, gypsum, anhydrite, halite, bitterns.

Non-marine evaporites are also typically dominated by halite, gypsum and anhydrite, although a wider range of minor salts do occur. This is because the chemical compositions of the original waters vary considerably according to the composition of the rocks with which they interact. Such non-marine evaporites include trona, mirabilite, glauberite, borax, epsomite, thenardite, gaylussite and bloedite (**Table 11.1**).

Principal sedimentary characteristics

See photographs and figures in the relevant sections of Chapter 3 as well as at the end of this chapter.

Bedding

A wide range of bedding styles is possible, depending on the depositional environment as well as post-depositional diagenesis and diapirism. Bed thickness ranges from very thin intercalations, for example within a peritidal or fluvial–lacustrine succession, to very thick more-or-less structureless units, especially where post-depositional changes have removed original bedding traces. Regular, parallel, thin to medium-bedded, cyclic evaporites typify deep-water successions, whereas irregular and nodular evaporites are common in sabkha deposits.

Extensive stratal dissolution may lead to collapse of the overlying beds and development of a collapse breccia. These may be very thick, widespread units, with no clear bedding or structure and little trace of the original evaporite. They can present spongy-textured, irregular horizons with distinctive yellow colouration (from sulphur), known as rauwacke or cargneule units. The removed evaporite finds its way into veins and other secondary deposits. The 'beds' of fibrous gypsum (satin spar), often found associated with limestones or other facies, are actually bedding-parallel displacive (intrusive) veins.

Structures

Nodular evaporites occur as discrete masses within mudrocks and also more closely packed, with only thin, irregular stringers of sediment (chicken-wire structure) in-between. They are typical of supratidal (or sabkha) environments. Crystalline forms of evaporite minerals also occur as isolated euhedral crystals in the background sediment – typically mudrock – and as entirely crystalline beds. Some twinned gypsum crystals (selenite) can be >1m in length. These are typical of lagoonal and intertidal settings, where they are often associated with signs of periodic desiccation – polygonal cracks, megapolygons and deep wedge cracks, tepee structures and cavities – and with microbial–evaporite stromatolites. Parallel-lamination, cross-lamination, and wavy, anastomosing bedding/lamination all indicate shallow to deep-water evaporites. Graded turbidite and debrite structures are recognized in deep-water evaporites. Dissolution of evaporite minerals can result in a very vuggy rock or, where replacement has occurred, give rise to pseudomorphs of the original nodule or crystal.

Texture

Crystal size and shape varies considerably within evaporite successions. Bottom-growth gypsum crystals, on the floors of lakes, lagoons, and shallow restricted shelves, commonly grow vertically as clear, well-formed selenitic crystals. They may reach spectacular sizes, over 1m in length, sometimes with curved crystal faces giving a palmate or cauliflower-like effect. Crystals precipitated in the surface waters of evaporating lakes or seas are mostly very fine-grained ($10–100\mu m$).

However, because evaporites are particularly prone to diagenetic modifications, including dehydration/rehydration reactions, dissolution, cementation, recrystallization, replacement, and deformation, their original texture is often destroyed. A fine, crystalline, sugary textured material, known as alabastrine gypsum, is a common diagenetic replacement for anhydrite. Larger crystals growing in a finer groundmass are called porphyrotopic gypsum. Pseudomorphic replacements, by evaporite or other minerals such as quartz, will mirror the original crystal size. Resedimented evaporites, including debrites and turbidites, are unusual in retaining at least part of their original textures.

Composition

Evaporite rocks are largely composed of single evaporite minerals, with variable trace impurities. Gypsum ($CaSO_4.2H_2O$) is the

Table 11.1 Principal types of evaporites

Major types	Subtypes	Composition and nature	
Marine evaporites (subaqueous and subaerial–sabkha precipitation)			
Chlorides	Halite	NaCl	Rock salt
	Sylvite	KCl	
	Carnallite	$KMgCl_3.6H_2O$	
Sulphates	Langbeinite	$K_2Mg_2(SO_4)_3$	Potash salts or bitterns
	Polyhalite	$K_2Ca_2Mg(SO_4)_4.2H_2O$	
	Kainite	$KMg(SO_4)Cl.3H_2O$	
	Anhydrite	$CaSO_4$	
	Gypsum	$CaSO_4\ 2H_2O$	
	Kieserite	$MgSO_4.H_2O$	
Carbonates	Calcite	$CaCO_3$	Inorganic carbonates
	Magnesite	$MgCO_3$	
	Dolomite	$CaMg(CO_3)_2$	
Non-marine evaporites			
Chlorides	Halite	NaCl	
	Rinneite	$FeCl_2.NaCl.3KCl$	
Sulphates	Gypsum	$CaSO_4 2H_2O$	
	Anhydrite	$CaSO_4$	
	Epsomite	$MgSO_4.7H2O$	
	Mirabilite	$Na_2SO_4.10H_2O$	
	Thenardite	Na_2SO_4	
	Bloedite	$Na_2SO_4.MgSO_4.4H_2O$	
	Glauberite	$CaSO_4.Na_2SO_4$	
Carbonates	Natronite	$Na_2CO_3.10H_2O$	
	Trona	$NaHCO_3.Na_2CO_3.2H_2O$	
	Gaylussite	$Na_2CO_3.CaCO_3.5H_2O$	
Borates	Borax	$Na_2B_4O_5(OH)_48H_2O$	
Silicates	Magadiite	$NaSi_7O_{13}(OH)_3.3H_2O$	

dominant sulphate mineral in modern evaporites, but undergoes dehydration to anhydrite ($CaSO_4$) during burial diagenesis so that anhydrite dominates at depths >600m. Trace elements present (from 1–100 ppm) include Br, Sr, B, F and Si. Strontium can also be present in much larger amounts, in excess of 1000ppm. Other impurities include organic carbon, clay minerals, quartz, feldspar, and sulphur. Their presence can affect the usual white, cream, or light- grey colour of evaporites: darker shades of grey are caused by traces of organic carbon, pinkish hues by iron, and yellowish ones by sulphur.

> **FIELD TECHNIQUES**
> Nearly all evaporite minerals are relatively soft (1.5–3.5 on Moh's hardness scale). Most halides, gypsum, and several others are readily scratched by a fingernail. Some of the other sulphates are only slightly harder, easily scratched with a coin.

Occurrence

EVAPORITES make up a minor part of the Phanerozoic succession (<2%) but are environmentally and economically very significant. They also occur from Precambrian times, the older ones mainly now as pseudomorphs rather than the actual salt. Thick, widespread deposits appear in late Cambrian, Permo– Triassic, Jurassic, and Miocene rocks, with lesser accumulations in the Silurian, Devonian, and Eocene. They form typically under marginal marine conditions (e.g. coastal sabkhas), but also occur in shallow to deeper-basin marine settings and in various non-marine environments (e.g. fluvial, lacustrine, eolian). They indicate climatic conditions where evaporation has exceeded precipitation, in other words arid, semi-arid, and hot, dry climates.

Deeply buried evaporites are subject to significant deformation and mobility. They act as preferred horizons for tectonic thrust planes and, nearer the surface, as basal glide planes for slide events. The low density and crystalline structure of halite, especially, leads to its relative buoyancy and tendency towards flowage within the stratigraphic succession. Salt diapirs may intrude through hundreds of metres of overlying strata, in some cases piercing through the surface.

Field photographs

11.1 Very thick gypsum/ anhydrite succession; part of Messinian evaporite basin of the former Tethys region.
Late Miocene, Psematisimenos Quarry, S Cyprus.

11.2 Thick gypsum unit (approx. 3.5m) sandwiched between micrite/marl intervals (M); note large cauliflower-like growth of gypsum in upper half of unit, and base of second gypsum unit visible near top; shallow-marine to lagoonal setting.
Late Miocene, Sorbas Basin, SE Spain.

11.3 Large selenite crystals on bedding surface of gypsum unit; indicative of slow crystal growth in tranquil conditions.
Width of view 15cm.
Late Miocene, Pissouri Basin, S Cyprus.

11.4 Detail of carbonate–evaporite lamination in which gypsum has been replaced by quartz (grey euhedral crystals). Dark bands are organic-rich. Width of view 6cm.
Upper Jurassic, Dorset, S England.

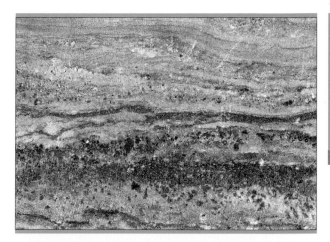

11.5 Finely laminated gypsum; note lenticular and micro-cross-lamination as clear evidence for current reworking of gypsum crystals; dark laminae slightly enriched in trace organics. Hammer 25cm.
Late Miocene, Psematisimenos, S Cyprus.

11.6 Laminated and cross-laminated gypsum bed, indicating current reworking of gypsum crystals. Hammer 25cm.
Late Miocene, Psematisimenos, S Cyprus.

11.7 Normally graded gypsum turbidite; note slight erosion at base, thin reverse-graded unit and highly fragmented gypsum clasts/grains.
Width of view 40cm.
Late Miocene, Pissouri Basin, S Cyprus.

11.8 Sabkha gypsum (irregular) within lake margin mudstone, as well as subaqueous gypsum layers (thin, white) as in **11.9**. Pen 15cm.
Mio–Pliocene, Lake Eyre, Australia.

11

11.9 Thin gypsum beds (white-coloured) in lacustrine mudstone succession. Hammer 25cm.
Pliocene, Lake Eyre, Australia.

11.10 Late development of displacive (intrusive) satin spar veins (white), in which the interlocking fibrous gypsum crystals grow at right angles to the overburden stress. Width of view 80cm. Photo by Ian West.
Cretaceous, Dorset, S England.

◄ 11.11 Enterolithic (contorted), secondary gypsum veins (centre of view), with secondary porphyrotopic gypsum (left), and rippled, sandy limestone (right). The gypsum developed after anhydrite from original primary gypsum. *Cretaceous, Dorset, S England.*

▶ 11.12 Bedding-plane view of pseudomorphs after cubic halite crystals in mudstone. Note characteristic hollow faces of cubes. Width of view 20cm. *Triassic, Nottinghamshire, UK.*

◄ 11.13 Dolomitized evaporite unit showing chicken-wire structure. This represents calcareous clayey seams around original nodular sabkha gypsum. Lens cap 6cm. *Cretaceous, near Benidorm, SE Spain.*

◄ 11.14 Dolomitized evaporite unit showing fibrous crystals over chicken-wire structure. Fibrous crystals represent calcite or dolomite replacement of original selenitic gypsum. Width of view 20cm. *Cretaceous, near Benidorm, SE Spain.*

▶ 11.15 Detailed view of halite and polyhalite, nodular and crystalline with red and clear patches. Soft (scratches with fingernail) and with strong, salty taste. Width of view 10cm. *Triassic, Saskatchewan, Canada.*

IRONSTONES

Brown-coloured metalliferous sediments (umbers) in the Troodos ophiolite, Cyprus.

Definition and range of types

IRON is one of the most common elements on Earth and is present in nearly all sedimentary rocks in minor amounts; a typical sandstone contains 2–4%, mudrock 5–6%, and limestone <1%. It is the presence of these iron-rich minerals that allow paleomagnetic measurements to be made, a magnetostratigraphic timescale to be determined, and the reconstruction of paleocontinental distribution to be attempted. It is also one of the principal colouring agents of sediments (see page 121). More rarely, iron minerals form a major component of the sediment, in which case the rocks are known as ironstones.

Ironstones are generally defined as a diverse suite of sedimentary rocks in which the iron content exceeds 15%.

They include:

- Thick, laterally extensive banded iron formations (BIF) that occur in Precambrian rocks throughout the world.
- Generally thinner Phanerozoic ironstones (PI) that are mainly marine/marginal in occurrence.
- Widespread deep-marine ferromanganese nodules, pavements, and crusts (FNod) that are mainly Recent and sub-recent in age.
- Thick, localized accumulations of metalliferous sediments (MSed) associated with mid-ocean ridge volcanism.

The only other modern environment where significant iron-rich sediments (bog-iron ores) are accumulating is in swamps and lakes of mid to high latitudes. A serious problem in the interpretation of ironstones is that none of the modern settings provides suitable analogues for ancient ironstones.

Within these major groups and spread of ages (Table 12.1), there are a variety of distinct ironstone facies associated with cherts, limestones, sandstones, mudstones, and volcaniclastic successions. The iron minerals present depend on the physico–chemical conditions that pertained both in the depositional environment and during diagenesis.

Principal sedimentary characteristics

See photographs and figures in the relevant sections of Chapter 3 as well as at the end of this chapter.

IRONSTONES are generally recognized in the field on the basis of their heavy weight (high specific gravity) and distinctive colour, which is mostly red, brown, and brown-black, more rarely with yellowish or greenish hues. However, not all ironstones are especially heavy (e.g. glauconitic sandstones, and metalliferous sediments), and their spectrum of colours is also found in many other sediments with only minor amounts of iron. Use of a hand lens can help estimate the percentage and types of iron minerals present, but further laboratory analyses are often also necessary. The main types identified above show both unique and overlapping characteristics. For convenience, they are referred to here by their abbreviations BIF, PI, FNod and MSed.

Bedding
- **BIF:** Thin-bedded/laminated, alternating Fe-rich and chert/quartz silt layers are characteristic of many BIF sediments. Lamination occurs at the microscale (0.2–2mm) and mesoscale (10–50mm), and some mesobands can be traced over thousands of square kilometers. Medium to thick and regular to irregular bedding, with features similar to those of PI sediments, also occur.
- **PI:** Bedding characteristics are very varied and reflect those of the sandstone, mudstone, limestone, or other facies that are iron-enriched.
- **FNod:** Nodular, irregular encrustation of bedding surface, and chimney-stack accretion forms are common; micro-lamination with irregular form also occurs in some deep-sea sediments.
- **MSed:** Thin to thick, well-bedded successions are common; more rarely with less regular or indistinct bedding.

Structures
- **BIF:** Finely laminated – parallel, graded, variety of microstructures common in fine-grained turbidites; also typical limestone-type facies with cross-bedding, desiccation cracks, stromatolites.
- **PI:** Wide range of types and features typical of sandstone, mudstone and limestone facies; including lamination, cross-lamination, grading, and bioturbation.
- **FNod:** Irregular encrustation lamination formed by mineral precipitation and growth.
- **MSed:** Structures generally absent.

Textures
- **BIF** and **PI:** Mainly sand, silt, and mud grades; some coarser-grained bioclastic conglomerates.
- **FNod:** Crystalline.
- **MSed:** Very fine-grained (mud-grade), with high microporosity – sticks to the tongue!

Composition
A wide variety of iron minerals is known from ironstones (Table 12.1). These occur either alone (as in nodules) or, more commonly, in association with the siliciclastic, carbonate, or volcaniclastic components of the host sediment. They are mostly dark-coloured grains within the sediment, that create an overall red, brown, black, or green hue.

Table 12.1 Principal types of ironstones and iron-rich sediments

Major types	Subtypes	Nature and origin
Precambrian Banded Iron Formations (BIF)	Algoma type	Relatively small lenticular iron deposits, associated with volcanics and turbidites. **Fe:** Hem, Gr, Sid, Py, Mag
	Superior type	Relatively large, thick, extensive iron deposits across stable shelves and in broad basins. **Fe:** Hem, Gr, Sid, Py, Mag
Phanerozoic Ironstones (PI)	Oolitic ironstones	Generally shallow marine, high sea-level, low sedimentation, ± condensed, oxygen-rich systems. **Fe:** Hem, Ch (Paleozoic) Be, Gt (Mesozoic) + minor Gl, Mag, Sid
	Bioclastic ironstones	Generally shallow marine, limestone facies, from open shelf to restricted coastal settings. **Fe:** Hem, Ch (Paleozoic) Be, Gt (Mesozoic)
	Glauconitic sandstones	Marine shelf, slightly reducing environment. **Fe:** Gl
	Iron-rich mudstones	Marine /lacustrine, quiet oxygenated to reducing environments; Fe finely disseminated and nodular. **Fe:** Py (associated organic-rich muds) Sid (associated coal succession) Be–Ch (associated marine muds)
	Placer deposits	Concentration in thin layers and laminae, typically within fluvial and shoreline sands. **Fe:** Mag, Hem
	Bog iron ores	Oolitic, pisolitic and nodular facies of medium to high latitude lakes, marshes and swamps. **Fe:** amorphous + Gt, Sid, and Mn oxides
Ferromanganese nodules, pavements, and metalliferous sediments (FNod, Msed)	Fe–Mn nodules and pavements	Mainly deep marine, oxygenated, bottom-current swept, low sedimentation settings. Also some shallow marine and lacustrine. **Fe:** Fe/Mn oxides + hydroxides + variety of accessory metals
	Metalliferous sediments (include umbers)	Mainly deep marine, spreading ridge settings, fluid emanations from black smokers. **Fe:** Fe/Mn oxides + hydroxides + high concentration of other metals
	Black smokers	Chimney vents on spreading ridges. **Fe:** as above

Hem = hematite, Gr = greenalite, Sid = siderite, Py = pyrite, Mag = magnetite,
Ch = chamosite, Be = berthierine. Gt = goethite. Gl = glauconite

Occurrence

See also Table 12.1.

BANDED IRON formations occur in the Precambrian cratonic cores of most continents. Two types are recognized on the basis of Canadian examples:

- *Algoma types* are mostly Archean in age (2500–3000 Ma), smaller-scale deposits, commonly associated with volcaniclastic rocks and/or turbidites in greenstone belts.
- *Superior types* are early–mid Proterozoic in age (1900–2500 Ma), generally thicker and regionally extensive, and deposited across stable continental shelves and in broad basins.

Several types or facies of Phanerozoic ironstone are recognized from a range of different environments and ages. Oolitic and bioclastic ironstones are typically shallow marine, normally oxygenated sediments, formed anywhere from open shelf to protected coastal settings. Glauconitic ironstones tend to form in offshore to outer shelf, slightly reducing marine environments. Iron-rich mudstones

> FIELD TECHNIQUES
> Look carefully for any indication of depositional environment on the basis of fossil content, trace fossils and associated facies. Otherwise, treat ironstones like any other sediment facies.

occur in both marine and lacustrine settings, under quiet low-energy conditions. Placer deposits are coarser-grained, higher energy, lag sediments typical of fluvial and coastal sandstones.

Ferromanganese nodules, pavements, and crusts are widespread deep-marine deposits in areas of fully oxygenated, bottom-current swept conditions, and very low rates of sedimentation. Their preservation potential is very low, although they are known from limited occurrences of Devonian and Jurassic pelagic limestones in Europe, and Cretaceous red clays of Timor. Black smoker chimneys and associated metalliferous sediments occur today on active spreading ridges, especially along transform faults, and have been found in association with numerous ancient pillow lavas and ophiolite suites.

Field photographs

12.1 Ironstone (magnetite, dark grey) and chert bands (light coloured) in Banded Iron Formation (Algoma type); note lamination, micro-cross lamination, contorted lamination and lenticular bedding of probable turbidite origin. Scale disk 2cm. Photo by Bob Foster.
Precambrian (Archean), Perfume Hill Prospect, Kraipan greenstone belt, South Africa.

12.2 Ironstone (magnetite, dark grey) and chert bands (light coloured) in Banded Iron Formation (Algoma type), with ductile deformation evident. Lens cap 6cm.
Photo by Bob Foster.
Precambrian (Archean), Perfume Hill Prospect, Kraipan greenstone belt, South Africa.

12.3 Ironstone (magnetite, dark), chert bands (light coloured), and siderite (brown) in Banded Iron Formation (Algoma type).
Photo by Bob Foster.
Precambrian (Archean), Lennox Mine, Mashava greenstone belt, central Zimbabwe.

12.4 Detail of **12.3** showing maculose (spotted) banded oxides. Photo by Bob Foster. *Precambrian (Archean), Lennox Mine, Mashava greenstone belt, central Zimbabwe.*

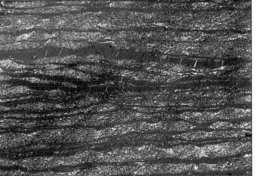

12.5 Banded Iron Formation (Algoma type) with evidence of transposed bedding and ductile deformation.
Photo by Bob Foster.
Precambrian (Archean), Athens Mine, Mvuma greenstone belt, central Zimbabwe.

12.6 Detail of Banded Iron Formation (Superior type). White-coloured, eye-like lenses are chert; brown-coloured layers are more ferruginous chert. Photo by Bob Foster. *Precambrian, Raposos Gold Mine, Minas Gerais, Brazil.*

12.7 Detail of Banded Iron Formation (Superior type). Red coloured hematitic layers alternate with laminated and micro-cross laminated quartz silt layers. Width of view 6cm. Photo by Bob Nesbitt. *Precambrian, Hammersley, Australia.*

12.8 Banded Iron Formation, with pyrite (bright specks) replacing banded jasper (red). Jasper is a red-coloured chert, its colouration due to finely disseminated hematite. Note also veins, fractures, and maicrofaults. Jasper interlaminated with other iron-rich layers is known as jaspilite, a common BIF sediment. Width of view 10cm. Photo by Bob Foster. *Precambrian, Adams Mount, Zimbabwe.*

12.9 Laterite; nodular, red-coloured, within highly leached paleosol horizon. Laterites are rich in insoluble iron and aluminium oxides, left as residual deposits after tropical weathering. Width of view 30cm.
Permian, Hallett Cove, S Australia.

12.10 Limonitic gossan (red-yellow iron-rich soil) over metalliferous sediments – possibly a sulphide facies of Banded Iron Formation in medium to high-grade metamorphic terrane.
Photo by Bob Foster.
Late Proterozoic, Chipirinyuma, eastern Zambia.

12.11 Oolitic ironstone (mainly hematite), forming a condensed horizon within limestone succession. The dark colouration is due to MnO_2 surface weather coating.
Lens cap 6cm.
Silurian, Quebrada Ancha, NW Argentina.

◀ **12.12** Oolitic ironstone (as 12.11), with desert varnish coating. Lens cap 6cm. *Silurian, Quebrada Ancha, NW Argentina.*

▶ **12.13** Detail of berthierine oolite (greenish), with well developed secondary siderite (brownish and partly collo-form). Width of view 6cm. *Jurassic, Northamptonshire, UK.*

◀ **12.14** Oolitic ironstone, now weathered to brownish limonite–goethite; associated with iron-rich siltstone/sand-stone. Trowel 25cm. *Cretaceous, Compton Bay, Isle of Wight, S England.*

◀ **12.15** Oolitic ironstone, now weathered to brownish limonite–goethite; lamination and cross-lamination locally evident. Trowel 25cm. *Cretaceous, Compton Bay, Isle of Wight, S England.*

▶ **12.16** Large, brown, sideritic ironstone nodules, fallen from cliffs and partially covered by sand. Nodules comprise iron carbonate (siderite), together with some terrigenous sediment, shells, and rare sharks' teeth. *Eocene, Hengistbury Head, S England.*

12.17 Ancient ferromanganese crust on surface of pelagic micrite bed; mainly iron and manganese oxides and hydroxides, coating and permeating the seafloor sediment. Oblique bedding-plane view; cross-section of bed in lower third of photograph.
Width of view 1.5m.
Paleogene, Mascarat, SE Spain.

12.18 Detail of bed (cross-section) in **12.17**, showing thin, darker brown ferromanganese crust over bioturbation and burrowing into micritic limestone.
Width of view 35cm.
Paleogene, Mascarat, SE Spain.

12.19 Core section split lengthways, showing modern, iron-(hematite) rich turbidite mud. Although the sediment is very red, the total iron content of 5–6% means it falls short of being classified as an ironstone.
Width of core (i.e. depth of frame) 8cm, top is to right.
Recent, Laurentian Fan, NW Atlantic Ocean.

12.20 Part of modern vent field showing broken and *in situ* vent chimneys and yellowish-brown metalliferous sediment on the seafloor; note also swarm of white, shrimp-like organisms. Width of view 3m, from submersible.
Photo by Bram Murton.
Recent, Broken Spur Vent Field, Mid-Atlantic Ridge.

12.21 Umbers, found in localized pockets over ocean crust pillow basalts of the Troodos ophiolite, and below pelagic chalks. Mainly iron and manganese oxides and hydroxides. Very fine grained and highly porous rock, quite light in weight; sticks to the tongue! Individual beds are completely structureless. For general view of setting between seafloor basalts and oceanic oozes, see chapter opener. Lens cap 6cm.
Late Cretaceous, near Pareklisha, S Cyprus.

SOILS, PALEOSOLS, AND DURICRUSTS

Calcrete surface crust and irregular caliche nodules below surface, Pissouri, S Cyprus.

Definition and range of types

SOIL is the thin residual layer developed upon solid bedrock and unconsolidated sediments through the action of physico–chemical and biological processes (pedogenesis). Soil type is dependent principally on climate, vegetation and rock type (Table 13.1). Thin, poor-quality soil develops at high latitudes as a result of physical weathering (entisols, inceptisols). More widespread are the soils of temperate climates (spodosols, podzols, and alfisols). Thick, rich soils develop in warm, humid areas (oxisols). Extensive leaching can result in the formation of hard nodular layers of iron and aluminium oxides/hydroxides (laterite and bauxite soils). High rates of

evaporation in arid and semi-arid areas leads to illuviation and precipitation of $CaCO_3$ nodules and layers within the soil (aridisols, mollisols). Silica, gypsum, and iron oxides may also occur.

Duricrusts refer to the hard coatings of these minerals, known as calcrete, silcrete, gypcrete, ferricrete, and so on, that develop either over the rock or soil surface, or as a hard cemented layer several tens of centimetres within the soil profile.

Paleosols are ancient soils and duricrust layers preserved in the rock record. Calcrete paleosols are recognized from the Precambrian onwards, although most of the older ones are now dolomitic. Ancient laterites and bauxites are also well known. Since the advent of land plants in Siluro–Devonian times, many of the other types of soil formed. Clayey soils formed beneath coal seams, typically leached of minerals and with marked rootlet traces, are known as seat-earths. Leached sandy soils are known as ganister.

Principal sedimentary characteristics

See photographs and figures in the relevant sections of Chapter 3 as well as at the end of this chapter.

THE RECOGNITION of paleosol horizons in the rock record is clearly of great paleoenvironmental significance and immediately demonstrates a su nvironment and processes. Paleosol type, including clay minerals present (on laboratory analysis), can help elucidate past climatic conditions.

Equally significant in field geology is the correct identification of modern soils and duricrusts overlying and distinct from older sediments. Extensive calcretization, characteristic of Mediterranean lands and other semi-arid climates for example, can completely obscure the true nature of sedimentary rocks exposed at the surface.

Soil horizons

Many soils possess four distinct horizons (*Fig. 13.1*): a surficial layer of organic matter (horizon O); a typically dark, fine nutrient-rich layer composed of humus, clays, and other minerals (horizon A or topsoil); a paler, coarser-grained layer rich in minerals precipitated out from groundwater leaching (horizon B or subsoil); and a layer of broken-up bedrock (horizon C or regolith). Soil profiles vary from this norm in terms of composition and completeness.

Clayey paleosols

Clayey paleosols (seat-earths) typically show colour mottling (especially reds, browns, and yellows), blocky textures, rootlets, rhizocretions (i.e. rootlet encrustations), and various

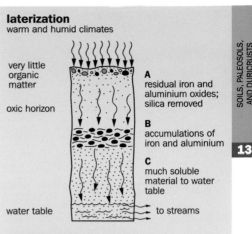

laterization
warm and humid climates

very little organic matter

oxic horizon

A
residual iron and aluminium oxides; silica removed

B
accumulations of iron and aluminium

C
much soluble material to water table

water table → to streams

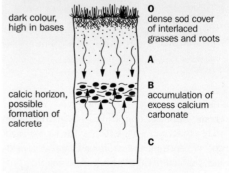

podzolization
cool and moist climate

humus rich

albic horizon

sandy, bleached

spodic horizon

low pH soil solution

water table

O
acid organic litter

A
eluviation of bases, oxides, clays

B
illuviation of oxides and clays

C
loss of bases to water table

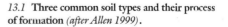

calcification
potential evapotranspiration equal to or greater than precipitation

dark colour, high in bases

calcic horizon, possible formation of calcrete

O
dense sod cover of interlaced grasses and roots

A

B
accumulation of excess calcium carbonate

C

13.1 **Three common soil types and their process of formation** *(after Allen 1999).*

Table 13.1 Principal types of soils, paleosols, and duricrusts

Major types	Subtypes	Nature and origin
Soils	Oxisols	Hot, humid, tropical soils; extensive leaching to laterites, bauxites.
	Aridisols	Hot, dry, desert soils; evapotranspiration and illuviation give duricrusts.
	Mollisols (chernozems)	Subhumid, semiarid grassland soils; organic-rich, well-developed, some duricrusts.
	Alfisols	Humid, temperate forest soils; grey-brown, moderately weathered.
	Ultisols	Subtropical forest soils; red-yellow, highly weathered.
	Spodosols (podzols)	Cool, moist climate soils of forests and heaths; brown-yellow, highly leached.
	Entisols	Embryonic soils with no developed profile.
	Inceptisols	Embryonic humic/sub-arctic soils.
	Vertisols	Tropical black soils with expanding clays.
	Histosols	Organic rich soils of bogs and peatlands.
	Andisols	Volcanic, chemically weathered soils.
Duricrusts	Calcrete (calcite)	Hard, cemented, crusts and layers formed where evaporation exceeds precipitation; although ferricretes (iron-pans) and alcretes can form where precipitation exceeds evaporation.
	Gypcrete	
	Silcrete	
	Ferricrete	
	Alcrete	
Paleosols (fossil soil horizons)	Sub-type less easily recognized	Soils with duricrust layers are best preserved.

nodules (e.g. siderite). Prolonged leaching of sandy paleosols may leave a massive quartz-rich sand layer (ganister), with or without colour mottling and rootlets. Extensive leaching characteristic of the humid tropics produces distinctive mottled yellow and red laterites and paler-coloured bauxites.

Calcretes

Calcretes are widespread as soils and a common paleosol type, forming especially in regions of the world with Mediterranean-type climates. They occur as nodules and layers with massive, laminated, and pisolitic textures. The characteristic fabric is a fine-grained, equigranular calcite mosaic that surrounds, replaces and displaces quartz or other grains and pebbles. Cracks, veins, spar-filled tubules (formerly rootlets), and rhizocretions are common. Microbial grain coatings, clusters, and crusts can all display a laminated structure.

Over prolonged periods of calcrete soil formation (from several to tens-of-thousands of years), scattered nodules develop into close-packed nodules and then to a massive limestone layer. Extensive calcite replacement may destroy all original texture and fabric.

FIELD TECHNIQUES

Be wary of what appears to be a widespread outcrop of limestone exposed at the surface, with no obvious bedding and often highly altered or weathered characteristics. This may be deeply calcretized sediment of completely different primary origin. Look carefully for ghost structures and components.

Field photographs

13.1 Deeply calcretized sediment, originally a dark green-grey, Pliocene-age, volcaniclastic-rich alluvial fan conglomerate, much of which is now replaced or obscured. Width of view 2.5m.
Recent, Melanda Beach, S Cyprus.

13.2 Deeply calcretized sediment, originally dark grey volcaniclastic sandstone. Width of view 5m.
Recent, Cabo de Gata, SE Spain.

13.3 Deeply calcretized sediment, originally dark coloured volcaniclastic conglomerate. Shoes – size 8 (42)!
Recent, Cabo de Gata, SE Spain.

13.4 Silcrete crust as thin, whitish layer over lacustrine siltstone. Width of block 15cm.
Recent, Lake Palankarina, Simpson Desert, Australia.

13.5 Reddened silcrete crust (hematite-stained) over wide area of desert surface.
Width of view 12m.
Recent, Cooper Creek, Simpson Desert, Australia.

13.6 Gypcrete crust and nodules within soil horizon. Soft, crumbly, and does not effervesce with HCl. Hammer 25cm.
Recent, Goyder, Lake Eyre, Australia.

13.7 Surface view of gypcrete crust with shrinkage cracks and loose pebbles of reddened silcrete. Hammer 25cm.
Recent, Goyder, Lake Eyre, Australia.

13.8 Gypcrete and calcrete crust, coating and permeating surface and sediment and replacing rootlet traces. Hammer 25cm.
Recent, Goyder, Lake Eyre, Australia.

13.9 Thick, red paleosol horizon between volcaniclastic-rich sandstone beds; fluvial to braid-delta succession. Width of view 5m.
Pleistocene, Pissouri Basin, S Cyprus.

13.10 Detail of thick paleosol horizon (as in 13.9) with well-developed caliche (carbonate) nodules. Hammer 30cm.
Pleistocene, Pissouri Basin, S Cyprus.

13.11 Thin paleosol horizon within alluvial–fluvial succession, with moderately well-developed caliche nodules. Wine pouch 15cm wide.
Pleistocene, near Benidorm, SE Spain.

13.12 Paleosol horizon with caliche nodules, above leached sandstone/pebbly sandstones within alluvial–fluvial succession. Width of view 2m.
Permo-Triassic, Raasay Isle, Scotland.

13.13 Ancient lateritic soil horizon, with nodular red-coloured laterite in highly leached, white-coloured siliceous/bauxitic material.
Permian, Hallett Cove, S Australia.

13.14 Modern soil – moderately well-developed podzol from weathering under temperate conditions; soil developed on 'brickearth', a glacial outwash/periglacial wind-blown loess deposit, overlying a flint gravel residual deposit from dissolution of underlying chalk with flints.
Recent, Lee-on-the-Solent, S England.

VOLCANICLASTIC SEDIMENTS

Rim and crater of Santorini volcano (now extinct), Aegean Sea. *Photo by Claire Ashford.*

Definition and range of types

VOLCANICLASTIC sediments are those composed mainly of grains and clasts derived from contemporaneous volcanic activity. They are a common part of many sedimentary successions and particularly dominant in active plate-tectonic settings. Some authors use the terms volcanogenic or pyroclastic as synonymous with volcaniclastic deposits (or volcaniclastites). However, the former implies a broader suite including extrusive volcanic rocks, and the latter is used here in a more restricted sense (see below).

Five major types of volcaniclastic sediment are distinguished on the basis of their origin, i.e. the process by which the original magma was fragmented into volcanic particles or tephra (Table 14.1).

These include:

- **Autoclastites**, formed by autobrecciation of lavas as they flow;
- **Pyroclastic-fall deposits**, that result from tephra fall-out from explosive eruptions;
- **Pyroclastic-flow deposits**, that result from different types of flows of tephra mixed with magmatic gas, steam or water;
- **Hydroclastites**, in which fragmentation occurs through magma-water contact;
- **Epiclastites**, that occur due to the reworking of volcaniclastic material by the normal agents of weathering and transport.

During reworking, the volcaniclastic material typically becomes variously admixed with siliciclastic, bioclastic or other sedimentary particles, giving rise to hybrid rock types. Terms such as sandy or calcareous epiclastite can be applied to impure volcaniclastic sediments. Likewise, siliciclastic or carbonate sediments containing volcanic tephra can be described as tuffaceous, also pumiceous and scoriaceous (see later).

Principal sedimentary characteristics

See photographs and figures in the relevant sections of Chapter 3 as well as at the end of this chapter.

VOLCANICLASTIC sediments are a compositionally distinctive suite of sedimentary rocks, some with unique characteristics resulting from their volcanic-related origin, and with exactly parallel features to those found in

14

Table 14.1 Principal types of volcaniclastic sediments

Major types	Subtypes	Nature and origin
Autoclastic deposits	Clast-supported autoclastite (volcanic breccia)	Poorly-sorted, angular breccias formed by autobrecciation of lavas as they flow.
	Matrix (lava)-supported autoclastite (volcanic breccia)	
Pyroclastic fall deposits	Agglomerate (>64 mm, bombs – fluid ejecta)	Formed through fall-out of volcanic fragments (tephra) ejected from a vent or fissure; can fall through air and/or water.
	Pyroclastic breccia (>64 mm, blocks – solid ejecta)	
	Lapillistone (2–64 mm, volcanic conglomerate)	
	Pyroclastic sandstone (0.06mm–2mm, coarse ash/tuff)	
	Pyroclastic mudstone (<0.06 mm, fine ash or tuff)	
Pyroclastic flow deposits	Ignimbrite flow	Formed from hot dense laminar flow of volcanic debris and magmatic gases.
	Surge deposit	Formed from dilute turbulent flow of volcanic debris and magmatic gas or steam.
Hydroclastic deposits	Hyaloclastites (non-explosive)	Formed by lava fragmentation through contact with water; subaqueous or subglacial eruptions and/or lavas flowing into water.
	Hyalotuffs (explosive)	
Epiclastic deposits	Epiclastic conglomerate, epiclastic sandstone, epiclastic mudstone, etc.	Formed by the reworking of any volcaniclastic material by the normal agents of winds, waves, currents, gravity flows, etc.
	Lahar deposit (volcaniclastic debrite)	Formed from either cold or hot, subaerial or subaqueous volcaniclastic debris flows.

siliciclastic sediments. They are often inter-bedded with lava flows and intruded by sills and dykes, and distinction from these igneous rocks is not always clear. They are also very susceptible to weathering and erosion so that epiclastites are a common part of any vol-canic–volcaniclastic succession, especially in more distal locations. Their relative chemical instability leads to rapid and extensive diage-netic alteration – devitrifying glass, altering minerals, creating a clay-rich matrix, and destroying depositional textures and struc-tures. Any field study should bear in mind these problems.

Bedding

A wide range of bedding styles is possible, ranging from very extensive, thin-bedded, ash-fall deposits, to laterally-restricted, very thick-bedded, wedge-shaped deposits. Bed geometry can help distinguish between fall, flow and surge types. Thick, composite units are typical of ignimbrite associations; whereas thin irregular basal breccias are common as autoclastites beneath lava flows (see *Fig. 14.1*).

Structures

There are many structures in volcaniclastic sediments that indicate their origin. Auto-clastic breccias have a chaotic, non-specific structure of jumbled clasts commonly above and below flow-banded lavas, or closely inter-mixed with the lava itself. Pyroclastic fall deposits typically show normal grading, although pumice clasts may display reverse grading particularly in water-lain beds. Impact structures caused by bombs (bomb sags) and large clasts are common, and welding can occur close to the vent. Ignimbrites are generally homogeneous in appearance apart from normal grading of larger lithic clasts and reverse grading of pumice clasts. Welding of individual grains to form a dense rock is typical of the hot middle parts of thick flows, together with flow-stretched pumice and glass (fiamme). Flow

stratification (parallel and cross-lamination; also antidune cross-bedding) is most com-mon in surge deposits, but may also occur in distal parts of ignimbrites. Hydroclastic deposits are generally structureless and chaotic, although lava flow over and into wet mud forms a distinctive rock known as peperite, showing an intimate and chaotic admixture of baked mudstone and chilled lava fragments or flow filaments. Epiclastic deposits show the full range of structures observed in conglomerates, sandstones and mudstones (see Chapters 4, 5 and 6). Par-ticularly common in proximal areas are lahar deposits, the result of volcanic debris flows known as lahars. These are generally struc-tureless, poorly sorted, matrix supported, and may show crude reverse grading with larger pumice clasts floating in the matrix.

Texture

Grain-size characteristics typically range from very coarse and poorly-sorted (including large boulders) close to source to extremely fine-grained and well-sorted in distal ash-fall deposits (Table 14.1, *Fig.* 14.2). Welded tuffs and ignimbrites display a crystalline texture, often difficult to distinguish from that of vol-canic igneous rocks. Grain shapes are com-monly angular and more or less equant, but also include euhedral–subhedral crystal shapes, well-rounded accretionary lapilli, and elongate fiamme. Volcanic grains in epiclas-tic deposits show more rounding with dis-tance from source and degree of reworking. Lahar deposits are very poorly sorted and can include large pumice blocks in a fine matrix.

Composition

Different types of magma – basic, intermedi-ate, acidic – are involved in the generation of volcaniclastic deposits, so that the particulate

14.1 **Bedding types and sedimentary structures in volcaniclastic deposits (synthesized from original sources).**

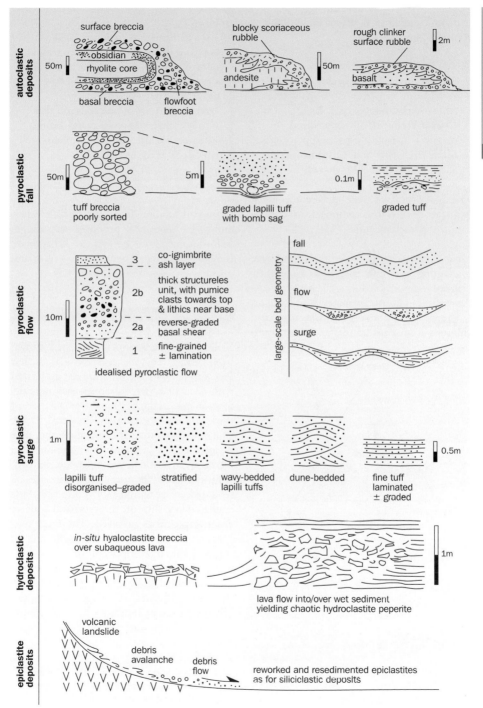

material or tephra will reflect these different compositions.

Tephra include:
- *Glass* from the liquid magma – highly vesicular pumice from more acidic lavas and scoria from more basic lavas.
- *Crystals* from partially crystallized magma – quartz, feldspar and pyroxenes.
- *Lithic fragments* from earlier eruptions and from the host country rock.

The proportion of these different components can be used for a compositional classification of pyroclastic tuffs.

Most older volcaniclastic sediments will have undergone significant alteration from their original composition during diagenesis. Glass devitrifies rapidly to clay minerals and zeolites – greenish (or brownish) smectites and chlorites from basic ashes, and paler, buff-

> **FIELD TECHNIQUES**
> Look for fiamme, accretionary lapilli, bomb sag structures, grading (reverse and normal), though many volcaniclastites are structureless and chaotic. Welded flow deposits can be confused with lavas. Alteration of unstable minerals and glass rapidly changes composition and texture. Pale greenish, lilac, and orange colours are typical of altered volcaniclastites.

coloured kaolinites from more acidic ashes. Submarine alteration produces an orange-yellow mineraloid known as palagonite. Replacement and cementation by silica and calcite is also common (See *Fig. 14.2*).

Occurrence

VOLCANICLASTIC sediments are well known from rocks of all ages, from earliest Precambrian to Recent. They are associated with areas of active volcanism and typically form an important, or even dominant, part of thick volcanic successions. Proximal to the vent or fissure source of an eruption, they may be interbedded with lava flows and intruded by sills and dykes. The unit thickness, individual bed thicknesses, and tephra clast sizes all tend to decrease within a short distance of the source, depending on the type and scale of eruptive processes involved. The finest ash, however, can be dispersed globally and so forms a minor component of other sediments. In general terms, autoclastites are the most proximal, followed by pyroclastic fall then flow deposits, and finally epiclastites. Hydroclastites are indicative of lavas coming into contact with water, wet sediment, or ice.

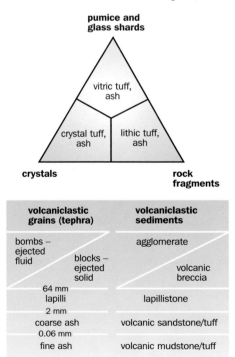

volcaniclastic grains (tephra)		volcaniclastic sediments	
bombs – ejected fluid	blocks – ejected solid	agglomerate	volcanic breccia
64 mm			
lapilli		lapillistone	
2 mm			
coarse ash		volcanic sandstone/tuff	
0.06 mm			
fine ash		volcanic mudstone/tuff	

14.2 **Classification of volcaniclastic grains and sediments based on grain size** (*below*)**, and pyroclastic tuffs based on composition** (*above*)**.**

Field photographs

14.1 Mixed volcaniclastic succession.
Quaternary, Pico de Teide, Tenerife, Spain.

14.2 Autoclastic breccia, blocky fragmented snout to left of subaerial lava flow (L), closely associated with co-eruptive and post-eruptive pyroclastic fall and flow deposits.
Width of view 3m.
Quaternary, Tenerife, Spain.

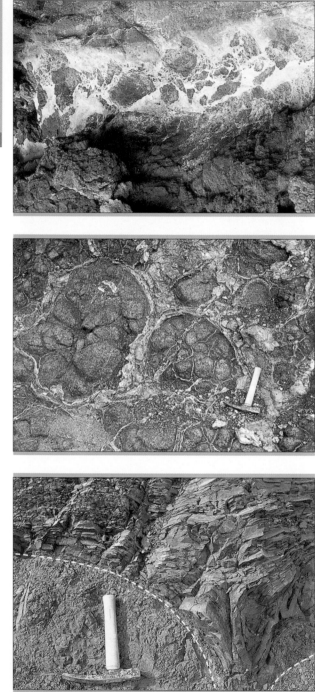

14.3 Autoclastic breccia encased in hydroclastite; dispersed, blocky, fragmented lava flow clasts showing quench fragmentation; white matrix is highly altered quench glass. Width of view 2m.
Quaternary, Tenerife, Spain.

14.4 Hyaloclastite quenched glassy rims (pinkish-white) round pillow lava; now undergoing devitrification and mineral alteration. Hammer 25cm.
Late Cretaceous, Troodos ophiolite, Margi, Cyprus.

14.5 Thin-bedded volcaniclastic turbidites over hyaloclastite pillow lava rims (dashed lines). Hammer 25cm.
Late Cretaceous, Troodos ophiolite, Margi, Cyprus.

14.6 Lava/wet-sediment inter-action – relatively thin andesitic lava flow (lighter colour) into lake sediments (darker grey), showing intense disruption of both lava flow and host sedi-ments, including sediment clasts incorporated into the flow and peperitic flow margins.
Hammer 40cm.
Triassic, Pichidangui, west central Chile.

14.7 Peperite, with dark coloured, fractured and angular, volcanic fragments in pale brownish sandy matrix.
Lava intruded into or over wet sediment. Width of view 40cm.
Neogene, Carboneras, SE Spain.

14.8 Peperite – as above, showing detail of highly irregu-lar lava–sediment interaction. Width of view 12cm.
Neogene, Carboneras, SE Spain.

14.9 Pyroclastic mudflow deposits, massive and chaotic volcanic breccia (agglomerate). Photo by Peter Baker. Width of view 10m.
Quaternary, Reunion Island, Indian Ocean.

14.10 Pyroclastic airfall and pyroclastic subaerial flow deposits, tuffs and lapillistone. Width of view 30m.
Quaternary, St Kitts, Caribbean.

14.11 Pyroclastic airfall deposit, trachybasaltic lapilli-stone; bedding-plane view. Cigarette 8cm.
Ordovician, Lake District, NW England.

14.12 Pyroclastic airfall deposit, andesitic lapillistone. Lens cap 6cm.
Miocene, near San Juan, Argentina.

14.13 Pyroclastic airfall deposits and pyroclastic mudflow deposits (lahars); interbedded agglomerate and lapillistone. Most prominent lahar is centre of view.
Quaternary, Tenerife, Spain.

14.14 Pyroclastic air-to-water tuff (white) and scoriaceous lapillistone (black). The lamination of the white tuff suggests some current reworking after deposition; the grading and sinking of larger clasts into underlying sediment suggests direct fall through water.
Lens cap 6cm.
Miocene, Miura Basin, near Tokyo, Japan.

14.15 Bedding plane view of pyroclastic air-fall to water-fall volcanic ash, fine-grained, pale coloured and light in weight (compared with heavier micrite, which can look similar). Note part of freshwater fossil fish, and scattered fragments of fossil leaves and insects.
Width of view 15cm.
Miocene, near Ankara, Turkey.

14.16 Pyroclastic air-to-water fall scoriaceous lapillistone/ volcanic breccia (black). Note marked normal grading, penetration of larger clasts into underlying hemipelagite, and probable current reworking at top of bed. Hammer 25cm.
Miocene, Miura Basin, near Tokyo, Japan.

14.17 Pyroclastic flow or debris avalanche deposit into and over lacustrine sediments (visible at base); note some dark-coloured lake sediment clasts have been incorporated within flow unit.
Hammer 30cm.
Triassic, Pichidangui, west central Chile.

14.18 Pyroclastic flow/debris avalanche unit – detail from 14.17. Note slightly wavy and sub-horizontal flow banding to right of flower.
Width of view 40cm.
Triassic, Pichidangui, west central Chile.

14.19 Pyroclastic flow deposit with flow banding now vertically oriented; top to left. Prickly pear cactus leaves approx. 15cm wide.
Neogene, San Jose, Cabo de Gata, SE Spain.

14.20 Pyroclastic flow deposit with streaked-out fiamme (dark coloured) and blocks. Fiamme are caused by the flow stretching of fragments of pumice and glass – in this case, dark glass fragments. Indistinct flow banding is also evident in parts.
Coin 2.5cm.
Quaternary, Tenerife, Spain.

14.21 Pyroclastic flow deposit with streaked-out fiamme (pale pinkish-coloured). In this case, the fiamme are flow-stretched pumice fragments. Coin 2.5cm.
Quaternary, Tenerife, Spain.

14.22 Base of thick pyroclastic flow unit (upper part of view), erosive into top of graded and laminated pyroclastic surge deposit. Coin 2.5cm.
Quaternary, Tenerife, Spain.

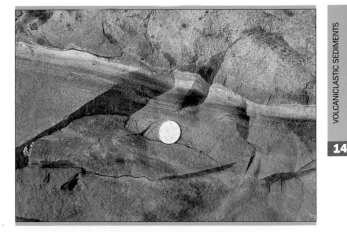

14.23 Pyroclastic flow deposit with partly weathered-out fiamme and fallen blocks. Coin 2.5cm.
Quaternary, Tenerife, Spain.

14.24 Pyroclastic surge deposit, showing detail of streaked-out gaseous lenses and parallel lamination. Coin 2.5cm.
Quaternary, Tenerife, Spain.

14.25 Pyroclastic mudflow deposit (or lahar), showing very thick indistinct bedding, and a matrix-supported fabric with scattered volcanic clasts in finer-grained volcanic mudstone. These have very similar features to those of true epiclastic debrites.
Photo by Paul Potter.
Tertiary, Sierra Madre Occidental Province, Chihuahua, Mexico.

14.26 Part of epiclastic mega-debrite deposit (lower 4m part of >30m thick bed), with crude stratification and reverse grading. Flow base is just above sea level; shoreline covered by fallen blocks.
Neogene, San Jose, Cabo de Gata, SE Spain.

14.27 Epiclastic, very thick-bedded debrite (part), showing indistinct parallel stratification, crude alignment of elongate clasts, and crude oscillation grading. Scale 30cm.
Carboniferous, Rio Tinto, S Spain.

14.28 Interbedded epiclastic turbidites (arrows, green) and hemipelagites (buff coloured). Note partial and complete Bouma structural sequences in the turbidites, and low-angle reverse faulting (right of hammer) due to tectonic compression. Hammer 25cm.
Mio–Pliocene, Miura Basin, near Tokyo, Japan.

14.29 Scoriaceous epiclastic sandstones with cross-lamination and minor shell debris; shallow-marine, possible tidal environment. Hammer 30cm.
Mio-Pliocene, Miura Basin, near Tokyo, Japan.

INTERPRETATIONS
AND DEPOSITIONAL ENVIRONMENTS

Puzzling over turbidites. Neogene basin fill deposits, Tabernas, Spain.

Building blocks

ANY STUDY of sedimentary rocks begins with the systematic observation of the sediments themselves. The many features and types of sediment observed in the field, as illustrated through Chapters 3–14, should be carefully recorded and used to construct a series of descriptive facies (see page 28). A sediment facies is defined as a sediment (or sedimentary rock) that displays distinctive physical, chemical, and/or biological characteristics (of the sort described in this book) that make it readily distinguished from associated facies in the locality. Facies include the field data collected on bedding nature, sedimentary structures, textural attributes, and composition.

Such data should first be collated and considered separately, as well as being combined into sediment facies. Both facies and individual data sets may be modified or refined through further sedimentological studies back in the laboratory.

Exactly the same approach is used for ancient sedimentary rocks in the field, for subsurface cored sediments, as well as for modern surface sediments. The facies described in each case form the building blocks for subsequent interpretation. These building blocks can be used to construct facies models, facies associations, cycles and sequences, lateral trends and geometry, and architectural elements – each of which is discussed briefly below.

There are a number of different aims in the study of sediments, both in the field and through subsequent analytical work. These include determination of:

- The source, flux, and rate of accumulation.
- The transport, depositional and post-depositional processes that have affected the sediment.
- Paleoclimate, climate change and their influence on the sedimentary record.
- Sea-level change through time.
- The tectonic control on sediment supply and sedimentary systems.
- The depositional environments and their evolution through the geological record.
- Lithostratigraphic units for geological mapping, perhaps as part of the broader structural/plate tectonic reconstruction of basin history.
- Economic evaluation for any one of a wide range of natural resources – oil and gas, metals, industrial minerals, water.

Whatever the ultimate aim of the study, geological interpretation is like very careful detective work. The picture is never clear, most of the evidence will always be missing, and the maximum amount of data must be painstakingly collected before any conclusion can be drawn. Laboratory work is often essential, as is the input of information from other disciplines – radioisotopic dating and micropaleontological analysis, for example. However,

> **IMPORTANT**
> Don't over-interpret your data or guess wildly from single observations. Usually it is only a careful combination of many different data that will allow firm interpretation. Consider all alternatives and try using field evidence to discriminate between them.

this book focuses on the primary sedimentary information that can be gathered in the field. Some pointers towards its interpretation are given in this chapter and some of the key texts to help with that interpretation are listed.

Facies characteristics and models

MUCH information on sediment source, as well as on the nature of transport and depositional processes, can be inferred directly from primary data collected on sedimentary structures, textures, and composition, as summarized in Chapter 3. Information on the sediment source is best gleaned from compositional data – synthesize and plot these on bar charts, pie diagrams, or cross plots. Clast types, sandstone grains, fossil fragments, and carbonaceous debris, for example, will all yield invaluable information. Consider variation of composition vertically through a succession as well as laterally across a sedimentary basin. Compositional and textural maturity will provide clues to the source proximality and transport distance.

Interpretation of sedimentary structures is one of the best and most direct ways of gaining information on depositional processes. Whereas lamination and cross-lamination, for example, are common to several different processes and environments, other structures can be more diagnostic – herringbone cross lamination, mud cracks, tepee structures, graded beds with structural sequences (Bouma, Stow, Lowe sequences), microbial lamination, fiamme, and accretionary lapilli, to list but a few. Read through Chapter 3 and carefully compare your field data with the photographs and figures given.

The primary data on sedimentary characteristics is still more powerful when synthesized into a series of descriptive facies for the study area. In the first instance, these are best kept relatively few in number and simple in nature – e.g. cross-laminated coarse sandstone, pebbly mudstone, black shale, and

15

Table 15.1 Sequence terminology, thickness, and timescale
The sediment facies approach and its comparison with sequence stratigraphic terms

Sediment facies
(= facies, lithofacies)

Sediment or sedimentary rock that displays distinctive physical, chemical and/or biological characteristics.

Descriptive facies Defined purely on sedimentary characteristics (e.g. muddy sandstone, laminated mudstone).
Genetic facies Determined by comparison with standard facies models and interpreted in terms of depositional process (e.g. turbidite, contourite, lahar deposit); commonly have standard sequence of structures through bed.
Thickness No absolute scale, typically 0.01–1m.
Timescale Very variable, geologically instantaneous (hours) for 1m thick sand-mud turbidite, or very slow accumulation (10–20ky) for 1m thick pelagite.
Sequence stratigraphy No equivalent. Do not confuse seismic facies, which have a completely different scale and interpretation.

Facies cycles
(small-scale sequences, microsequences, cyclothems, rhythms)

Rhythmic alternation of two or more facies, or small-scale systematic changes in bed thickness, grain size or other properties.

Thickness Typically 0.5–5m, generally <10m.
Timescale Repetition typically 10–100ky, may be Milankovitch cyclicity of 20ky, 40ky or 100ky approximately (= 4th and 5th order of sea level change); also more rapid or less regular.
Sequence stratigraphy No equivalent, although some of thicker facies cycles may be equivalent to thin parasequences.

Architectural elements

Large-scale 3D building blocks of depositional systems made up of several different facies and cycles, and one or more mesosequences. They represent specific subenvironments (e.g. channels, levees, lobes, mounds, etc).

Thickness Typically 5–100m, variable width/length ratios.
Timescale Construction over 100ky–1My (=3rd and 4th order of sea level change).
Sequence stratigraphy In part, equivalent to systems tracts or depositional units.

Facies associations
(large-scale sequences, macrosequences)

Complex arrangement of different facies, cycles, mesosequences, and elements grouped together in distinct units, characteristic of a particular depositional setting (e.g. submarine fan, delta-top, carbonate reef to bank system).

Thickness Typically 50–250m, variable areal extent.
Timescale Develops over 0.5–10My (=2nd and 3rd order of sea level change).
Sequence stratigraphy Equivalent to the fundamental unit in sequence stratigraphy, known as the sequence or depositional sequence

---

Basin-fill sequences
(megasequences)

Thick sedimentary succession comprising varied elements and facies associations, representing the complete fill of a sedimentary basin.

Thickness Typically 250–>1000m, variable areal extent.
Timescale Depends on basin size, ranges from <1–100My (1st–3rd orders).
Sequence stratigraphy Known as basin-fill succession.

INTERPRETATION

15

graded silt-laminated mudstone. For detailed work, it may then be necessary to divide single facies into several subfacies. It is also useful practice to compare descriptive facies from the study area with standard facies models for different depositional systems. These are idealized associations of lithology, structure, texture, and/or other features that can be interpreted in terms of the hydrodynamic process of deposition, and may be further linked with a typical depositional environment. Turbidite, debrite, and contourite facies models, for example, are typical of turbidity current, debris flow and bottom (contour) currents, respectively, in deep-water environments (see pages 56–66).

Key references: Tucker and Wright (1990), Walker and James (1992), Reading (1996), Leeder (1999). But see also basic sedimentology texts listed in bibliography

Facies sequences and cycles

THE NEXT STEP in analyzing field data is to consider the vertical stacking of individual facies, to ascertain whether or not there is some systematic or cyclic organization apparent. In some cases, the facies are interbedded more or less randomly. In other cases, there is a preferred order of vertical transition from one facies to another. These are known as facies sequences, and may occur at several different scales from <1m to several hundreds of metres. Table 15.1 illustrates the nature and scales of these different sequence types, the terminology used, and its comparison with

the rather different terminology developed in sequence stratigraphy (see below).

Small-scale sequences (microsequences; <10m thick, typically 0.5–5m) include: simple alternations of facies (e.g. chalk–chert, limestone–marl, sandstone–mudstone); the repetition of several different facies (e.g. micrite–grey, bioturbated marl–black shale, mud–mottled silt–sand); or the systematic oscillation of grain size, bed thickness, and/or other characteristics – such as compensation cycles in turbidite systems. These can be interpreted as a simple oscillation of climate or productivity, for example, or repeated turbidite input onto a basin plain. The repetition may be on a Milankovitch scale – 20, 40, or 100ky – or much more rapid and less regular.

Medium-scale sequences (mesosequences; typically 5–50m thick) show a systematic superposition of several different facies and/or bed thickness and grain size. Such sequences have been characterized from many different environments as the result of natural progressions and environmental change. Examples include delta, beach, or submarine-fan lobe progradation, lateral migration of a meandering river, channel-fill on a deep-sea fan. The repetition is generally between 100ky and 1My. Both small- and medium-scale sequences are also known as cycles, cyclothems or rhythms, so that some confusion of scale and time may result unless due care is taken. Some typical sequences at these scales are described in *Figs 15.1, 15.2* and illustrated in *Plates 15.1–15.11.*

sequence	type	occurrence/ interpretation
	limestone chert (flint) limestone chert (flint)	common in many deep-water carbonate systems; Milankovitch climate cycles influencing biogenic productivity
	marl limestone marl	common in many shelf and deep-water carbonate systems; Milankovitch climate cycles influencing biogenic productivity
	organic-poor bioturbated mud finely laminated black shale organic-poor bioturbated mud	common in black-shale systems (e.g. petroleum source rocks); cyclic input and/or preservation of organic matter, alternation of oxic/anoxic conditions at seafloor; climatic or other forcing mechanisms
	bioturbated mud graded-laminated silt-mud graded sand	common in many turbidite systems, especially medial–distal settings e.g. basin-plain deposits); episodic turbidity current input
	bioturbated mud mottled silt/mud (± indistinct lam.) bioturbated sand (± lens) mottled silt/mud bioturbated mud (± indistinct lam.)	common in many contourite systems, in slope – deep water settings; grain-size cycles due to long-term change in mean bottom-current velocity; climatic or other forcing mechanism
	sand pebbly sand gravel pebbly sand sand (± internal stratification throughout)	common in many high-energy siliciclastic systems – e.g. fluvial, alluvial, nearshore, turbidite; generally the result of fluctuation in flow velocity – either episodic or semi-continuous
	mud/marl ± volcaniclastic coarse-grained volcaniclastic bed ± graded	common in subaqueous volcaniclastic systems; episodic pyroclastic fall or epiclastic input (e.g. turbidites) into basin with background (hemipelagic) sedimentation

scale of sequences – typically decimetric

Several different techniques can be used to test a succession for facies relationships, to help determine the order or randomness of sequences. These include counting the number of times each facies is overlain by the others, measuring bed thickness or bed grain size, noting the nature of bed boundaries, and so on. Data can be presented as simple facies relationship diagrams and/or analysed statistically. An important statistical analysis commonly used is the Markov Chain, or its variants, as described in the key references. Larger-scale sequences, facies associations and architectural elements are discussed below.

Key references: Reading (1986, 1996); Lindholm (1987); Graham, in Tucker (1988), Walker and James (1992), De Boer and Smith (1994).

Lateral trends and geometry

(Plates 15.11–15.17)

THE HORIZONTAL or spatial arrangement of facies and distinct lateral trends in facies characteristics are also very important for interpretation. At the scale of a single exposure, significant but limited observations can be made – e.g. sand-body geometry, unconformity surfaces (with onlap, offlap or other relationships), and paleocurrent/paleoslope orientation.

By careful correlation of the same interval over a large area, further observations can be made. This can be achieved by walking out exposures, and by lithostratigraphic, biostratigraphic and chronostratigraphic correlation. Although accurate detailed work may require additional input or laboratory analysis, the first broad-brush correlations can

15.1 Common simple facies sequences in sedimentary successions, their occurrence and likely interpretation. Also called cycles and rhythmic bedding. Typical scale of one cycle or couplet is a few tens of centimetres (range few centimetres to few metres).

often be made in the field, followed by measurements of:

- Large-scale geometry – e.g. thickness variation of lithostratigraphic units.
- Relationships between lithostratigraphic units.
- Regional paleocurrent and paleoslope trends from a variety of facies and data types.
- Composition and grain size trends, and the lateral variation of compositional and textural maturities.
- Regional trends in facies types and sequence types.

These data can be interpreted in terms of depositional geometries, the location and trend of shoreline or slope, sediment source and distribution patterns, proximal–distal relationships, and architectural elements.

Architectural elements and facies associations

THE NEXT step in interpretation is to link together observations of small and medium-scale sequences, lateral trends, bounding surfaces between sequences, and any observed 3D geometry into architectural elements. These are the larger-scale features or building blocks that seek to incorporate both 2D and 3D data. In the deep-sea environment, for example, we can identify the following elements: hiatuses, erosional plains and boundary surfaces; marked gradient changes; erosional slide and slump scars; canyons and channels; levees, overbank, and interchannel regions; mounds and lobes; contourite drifts; sheets, drapes, and megabeds.

As an example, a deep-water channel element can be characterized by an erosive base, slumped margins, a sediment fill of debrites, coarse-grained and fine-grained turbidites arranged in a fining-upward sequence, a unidirectional to splayed paleocurrent

pattern, and downchannel decrease in grain size and bed thickness. Some of these features, as well as those for other elements/associations, are shown in *Plates* 15.33–15.63 for different depositional environments.

At the next level of organization, large-scale sequences (macrosequences; typically 50–250m thick) represent a long-term and complex arrangement of facies, sequences and elements. Several different facies, microsequences and mesosequences, lateral trends and geometry, as well as architectural elements (*Fig 15.4*) are commonly grouped together in distinct units, characteristic of a particular depositional setting and/or mode of formation. These are known as facies associations. Examples might include a complete carbonate bank–basin system, a submarine fan complex, a delta-top to pro-delta system, and so on. The duration of such sequences is generally of the order of 0.5–10My.

Basin-fill sequences (megasequences; typically 250m to >1000m thick) are the largest scale of sedimentary succession that can be identified where outcrop is particularly well exposed or readily correlated between sections. Although there may not be any regular pattern to basin fill, an overall fining-upward sequence related to tectonic uplift, denudation, and sediment supply is quite common.

Key references: Reading (1986, 1996); Friedman et al (1992), Walker and James (1992), Leeder (1999).

Sequence stratigraphy and bounding surfaces

(Fig 15.3, Fig 15.4)

SEQUENCE stratigraphy can be defined as the analysis of genetically related depositional units within a chronostratigraphic framework. It is a parallel system for interpreting sedimentary systems (to that outlined above) that has grown out of the subsurface analysis of continental margins by oil explorationists, using seismic profiling and deep borehole

a

mudrock, often red with calcareous nodules (calcrete): deposition by vertical accretion on floodplain

fining-up sandstone, cross-laminated, cross-bedded (mostly trough), flat-bedded: deposition on point bar

thin conglom. } channel
scoured surface } floor

mud | f'm'c sand

b

coal/soil horizon

sandstone: with cross-bedding (various types), flat bedding + channels

sandstone/mudrock with lenticular + wavy bedding

mudrock

mudrock or limestone with marine fossils

mud | f'm'c sand

c

back beach eolian dunes

beach sands

upper shore face

lower shore face
– bioturbated
– coarsens upwards

offshore (shelf) muds

mud | f'm'c sand

d

aeolian sand, large-scale cross-beds (dunes)

flat bedding in truncated sets, cross strat. (beach)

cross-stratified, flat-bedded sandstone (shoreface) HCS/SCS sandstone

mudrock, bioturbated, storm beds (open shelf)

mud | f'm'c sand

e

palaeokarstic surface sabkha evaporites,

microbial laminites (tidal flat) wacke/packstones restricted fauna (lagoonal)

oolitic/bio-grainstone cross strat.(shoal, barrier)

mud/packstone, rich fauna, graded storm beds (open shelf)

mud | w'p'g

f

paleosol ± calcrete

biopelmicrites/wacke-stones ± stromatolites (tidal flat)

biopel- + oo-sparites (intertidal–shallow, subtidal)

bioherms + biostromes

biopelmicrites/wacke-stones ± terrigenous clay, bioturbation, storm beds (deeper-water subtidal)

basal intraformational conglom. (transgression)

g

micrites, cherts (deep basin)

slide–slump units (slope instability)

calcarenite turbidites

calcarenite turbidites + pelagites (slope)

micrites, cherts (deep basin)

micrites, black shales (anoxic basin)

h

volcaniclastic surge + fall (emergent)

volc./shelly x-strat sandstone (shallowing up)

volcaniclastic debrite

volcaniclastic turbidites (slope)

hemipelagic marls + volcanic tuff (basin, slope)

i

enterolithic anhydrite

nodular/chicken-wire anhydrite

stromatolites ± gypsum pseudomorphs

fenestral limestones

pelleted limestones

oolitic, bioclastic limestones

marls, claystones

} supratidal
(sabkha) sediments

} intertidal

} shallow subtidal

} deeper-water subtidal

j

massive volcaniclastic sand + volc. slide/debrites (basin fill)

bioherm talus

calcarenite turbidites

alternating micrites + mudstones (deep-water)

k

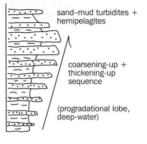

sand–mud turbidites + hemipelagites

coarsening-up + thickening-up sequence

(progradational lobe, deep-water)

l

sand–mud turbidites, debrites + hemipelagites

fining-up + thinning-up sequence

(channel-fill, deep-water)

techniques. From these datasets, the bounding surfaces between what are then interpreted as different depositional systems, are relatively easily identified on seismic profiles and dated in isolated borehole sections. Limited data are available on the actual sediments. Both bounding surfaces and intervening systems are related to global sea-level change.

The principal sedimentary body in sequence stratigraphy is the depositional sequence (or simply sequence), which is equivalent to the large-scale sequence or facies association (above). This is a complex body sandwiched between successive lowstand sequence boundaries, and itself made up of

15.2 Common complex facies sequences in sedimentary successions; typical thicknesses for each sequence in parentheses, below. Note there can be much variation from these typical sequences.

a) Fining-up sequence produced by lateral migration of a meandering stream (5–35m).
b) Coarsening-up sequence produced by delta progradation in its simplest form (5–50m).
c) Coarsening-up sequence produced by shoreline progradation over muddy shelf facies (10–50m).
d) Coarsening-up sequence produced by progradation of a beach-barrier system (10–50m).
e) Shallowing-up carbonate sequence, including mainly shallow marine carbonates (5–50m).
f) Shallowing-up carbonate sequence from submarine to subaerial exposure, including reef development (5–35 m).
g) Deep-water slope to basin facies association, with coarsening-up sequence of resedimented carbonate facies (10–50m).
h) Shallowing-up volcaniclastic sequence (20–100m).
i) Shallowing-up sabkha sequence (5–50m).
j) Volcanic margin slope sequence (20–100m).
k) Coarsening-up (thickening-up) to asymmetric turbidite sequence produced by lobe deposition and switching (5–60m).
l) Blocky to fining-up (thinning-up) turbidite sequence produced by submarine channel-fill and abandonment (10–100m).

Compiled from multiple sources, including Tucker 1991, 1996; Reading 1996; and Stow, various.

15.3 Log-plot comparing span of rates for (a) sedimentary accumulation and denudation, and (b) tectonic uplift, subsidence, plate motion, and sea level.
Compiled from various sources.

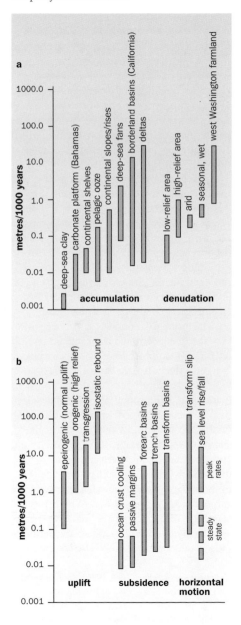

15.4 Typical geometries of individual beds and large-scale sediment bodies; the nature of different bounding surfaces and relationship between sedimentary units.

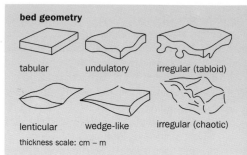

smaller-scale bodies known as system tracts. A complete sequence comprises, sequentially, a lowstand system tract, a transgressive system tract, a highstand system tract, and a regressive system tract. These are more or less equivalent to architectural elements (above). Where system tracts are composed of smaller-scale identifiable cycles, these are known as parasequences (or depositional units), which are equivalent to medium-scale sequences (above).

It is now, therefore, both common and useful practice to relate field sedimentology to a sequence stratigraphic framework. Field observations are then more directly comparable with subsurface data and more readily used as ancient analogues for hydrocarbon exploration. To do this, much hinges on accurate field identification of bounding surfaces and precise dating throughout the field area to ensure that such surfaces truly correlate.

The principal boundaries between individual sequences should be examined carefully for evidence of significant erosion, a marked reduction in the rate of sedimentation (e.g. intense bioturbation), a distinct break or hiatus (e.g. hardground or paleosol development), subaerial exposure and/or reworking of the top of the underlying sequence (e.g. sharp scoured surface, lag deposit), and abrupt facies change between facies of two distinct environments, and so on. The types of boundaries and their characteristics are summarized in *Fig 15.4*. Where several sequences or parasequences occur in a succession, then any upward change in the nature of individual sequences should also be noted.

Key references: Reading (1986, 1996); Friedman et al (1992), Posamentier et al (1993), Weimer and Posamentier (1993), De Boer and Smith (1994), Emery and Myers (1996), Miall (1997).

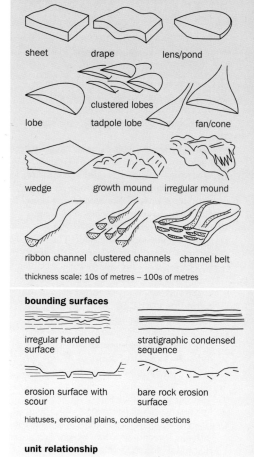

Cycles, geometry, and unconformities

15.1 Limestone–marl cycles. Cycles have a Milankovitch periodicity and probably represent climatic-induced variations in marine biogenic productivity. *Pliocene, Trubi Marls, Sicily.*

15.2 Limestone–black chert cycles. The black cherts are rich in organic carbon and increase in thickness and frequency towards the mid-part of this section. They display a complex Milankovitch cyclicity at the time of the mid-Cretaceous Ocean Anoxic Event, a period of time when many parts of the world's oceans were poised at very low oxygen levels. Width of view 12m. *Cretaceous, Scaglia Bianca, central Italy.*

15.3 Turbidite sandstone–mudstone couplets, showing a thickening-up to thinning-up symmetrical sequence, as marked (top to left). May represent a small isolated lobe deposit in a slope-apron succession. *Triassic, Los Molles, west central Chile.*

15.4 Turbidite basin-fill succession showing a large-scale coarsening-up sequence (about 300m thick), from dark mudstone to pale-coloured, thick-bedded sandstones.
Eocene, Annot Basin, SE France.

15.5 Limestone–marl succession showing thinning and drape over a small patch-reef complex (R). Note fault (F) with minor displacement to left of mound. Width of view 16m.
Miocene, Pakhna Formation, southern Cyprus.

15.6 Lenticular lens of conglomerates, probably debris-flow fill of small channel, part of alluvial–fluvial succession above angular unconformity (dashed line) with Cretaceous nodular limestone–marl succession.
Plio–Pleistocene, Sierra Helada, SE Spain.

15.7 Wedge-thinning of turbidite sandstone lenses (dashed lines); note also that the topographic low to right of lobe margin of the thick sandstones is filled by subsequent thin-bedded turbidites (a compensation effect). Width of view 5m. *Eocene turbidite basin, Annot, SE France.*

15.8 Turbidite compensation cycles in sandstone–mudstone turbidite succession showing thinning-up and thickening-up patterns of bed thickness (as marked). These are caused by each succeeding turbidity current responding to the very subtle relief of previously deposited beds, and so flowing just off the axis of maximum thickness. The result is a systematic variation in bed thickness, typically through a sequence of 3–7 beds. Width of view 3m. *Paleogene, S California, USA.*

15.9 Margin of irregular block-like channel fill (dashed line)–conglomeratic debrite over thick turbidite sandstone and slurry bed – cut into thin-bedded turbidite succession.
Miocene, Tabernas Basin, SE Spain.

15.10 Onlap, thinning, and pinchout of turbidite sandstones against submarine channel margin (dashed line) that shows draped beds with evidence of sliding (S). Width of view 8m.
Paleogene, S California, USA.

15.11 Shallow symmetrical channel filled with calcisiltites and calcarenites; probable tidal environment. Dashed line marks base of channel fill.
Photo by Paul Potter.
Carboniferous (Mississippian), Pulaski County, Kentucky, USA.

15.12 Large-scale sigmoid progradation geometry (dashed lines) at margin of carbonate platform. The horizontal topset beds are clearly seen overstepping the inclined slope beds. Height of cliff section approximately 50m.
Triassic, Dolomites, N Italy.

15.13 Angular unconformity (dashed line) with Plio–Pleistocene alluvial–fluvial succession overlying Cretaceous nodular limestone–marl succession.
Sierra Helada, SE Spain.

15.14 Apparent unconformity dashed line) below buff-coloured sandstone unit (Miocene turbidites) is probably due to recent landslide (LS). Disruption and apparent unconformity within the mudrock succession (Oligo–Miocene thin-bedded silt–mudstone turbidites) is interpreted as a post-depositional submarine slide–slump (SS) event; basal slip plane marked with dotted line. Width of view 25m.
Urbanian Basin, Umbro-Marche region, central Italy.

15.15 50m-high exposure through turbidite–hemipelagite succession of the Tabernas Basin. Apparent unconformities (dashed lines) are interpreted basal glide planes within a large-scale submarine slide, which has probably been displaced by some 5–10km from the north. The basal glide plane lies on a chaotic debrite unit (D).
Miocene, Tabernas Basin, SE Spain.

15.16 Unconformity without marked angular discordance (dashed line), but indicated by widespread erosion, intense bioturbation and a high degree of iron staining. Width of view 6m.
Late Pleistocene fluvial gravels over Eocene shallow-marine glauconitic sandstones, Lee-on-the-Solent, S England.

15.17 Minor unconformity (disconformity) within late Jurassic succession, marked with arrows; oolitic limestone above highly bioturbated and burrowed shallow marine sandstones. This irregular surface is marked by minor erosion, intense bioturbation, and iron staining; it also juxtaposes two very different sedimentary facies. Width of view 50m.
Osmington Mills, Dorset, S England.

15.18 Part of major strike–slip fault zone (>250m wide in places), showing ground-up interdigitated slivers of various rock types – the Rainbow Fault. Width of view 20m.
Neogen, Mojacar, SE Spain.

15.19 Closely spaced set of minor high-angle reverse faults through parallel-laminated volcaniclastic sandstone. Sense of movement indicated by arrows. Key tag 6cm.
Miocene, Carbonerus, SE Spain.

15.20 Bed-parallel fault surface coated with calcite, within silt-stone–mudstone succession. Note that the orientation of calcite crystals (lineation) indi-cates the orientation of fault movement (arrow). Steps in the fault plane futher indicate the sense of movement. Width of view 3m.
Precambrian, Hallett Cove, S Australia.

Sediment interpretation in cores

◀ **15.21** Mudstone and silty mudstone with bioturbational mottling.
Mid-Jurassic, North Sea oilfield.

▶ **15.22** Sandstone reservoir, with indistinct lamination/cross-lamination near base.
Mid-Jurassic, North Sea oilfield.

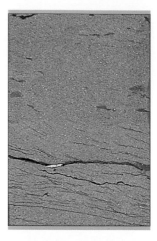

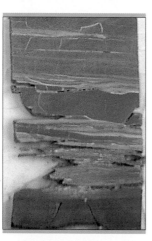

◀ **15.23** Fine-grained silt–mud turbidites with partial Stow sequences.
Paleogene, North Sea oilfield.

▶ **15.24** Medium-scale cross-laminated sandstone reservoir, with marked erosional surface near base.
Mid-Jurassic, North Sea oilfield.

◀ **15.25** Cross-laminated sandstone reservoir, with marked erosional scour. Note sample plugs for poroperm measurement.
Mid-Jurassic, North Sea oilfield.

▶ **15.26** Small-scale cross-lamination and parallel-lamination in shallow-marine/lagoonal silt–mud facies.
Mid-Jurassic, North Sea oilfield.

◄ 15.27 Small-scale (ripple) cross-laminated, fine-grained sandstone, with mudstone partings.
Mid-Jurassic, North Sea oilfield.

►15.28 Carbonaceous shale and thin coal horizon, below sandstone with rootlet traces (rootlets from overlying coal).
Mid-Jurassic, North Sea oilfield.

◄ 15.29 Interbedded silt and mud facies, with flaser to wavy lamination, bioturbation/ burrows, and small siderite concretion.
Mid-Jurassic, North Sea oilfield

►15.30 Bioturbated silt and mud facies, now indistinctly laminated, from interdistributary bay complex.
Mid-Jurassic, North Sea oilfield.

◄15.31 Bioturbated and laminated, carbonaceous shale over coal (bottom left). Note irregular pyrite concretion.
Mid-Jurassic, North Sea oilfield.

►15.32 Dark mudstone and siltstone over muddy sandstone, with bioturbation and burrow traces.
Mid-Jurassic, North Sea oilfield.

Sediment interpretation in cores

(*Plates 15.21–15.32*)

ALTHOUGH this book has focused on sediment description in the field and analysis of sedimentary rock outcrop data, exactly the same characteristics are evident in subsurface cores. Our primary knowledge of sediment facies and their distribution in modern environments at sea comes from studies of unconsolidated sediments that have been recovered from below the seafloor. A range of different coring devices can be used – gravity cores, piston cores, vibrocores, and box cores, for example – most of which yield narrow cylinders of soft sediment from the upper few metres of section.

Scientific drilling deeper into the sediment column has yielded much thicker successions (up to about 1500m) of subsurface cores, from the unconsolidated surficial sediment, through semi-consolidated to fully consolidated and cemented sedimentary rock. Borehole drilling into deeply buried sedimentary successions for economic resources of all types is generally accompanied by recovery of sediment cores from selected intervals of importance (up to several thousands of metres below the land surface or seafloor), as well as by remote sensing of the subsurface via wireline logging techniques. Whether for scientific purposes, for coal or mineral exploration, or for hydrocarbon exploration and production purposes, very similar thin cylinders of core are recovered. These are split lengthways into two halves and then examined in exactly the same way as for sedimentary rocks in the field, except that the 'exposure' observed is no more than a thin, vertical (inclined, in some boreholes) strip of rock.

Some examples of cored successions and associated sediment logs are shown in *Plates 15.21–15.32*. Also indicated are the outcrop features that can help with their interpretation. It is intended that this book can be used equally for comparison with, and interpretation of, cores.

Controls, rates and preservation

(*Fig. 15.3*)

FOR ANY interpretation of sedimentary systems it is important to consider the principal factors that have influenced the accumulation of sediment and its preservation. These include both external or allogenic controls and internal or autogenic controls.

Allogenic controls
- *Sediment supply*: the nature, rate, and source of supply, as well as the type of sediment.
- *Climate*: temperature, precipitation, and wind regimes; also short/long-term climatic change due to Milankovitch cycles.
- *Tectonic activity*: isostatic movements, subsidence and uplift, plate tectonic setting, seismicity, and volcanicity.
- *Sea-level changes*: eustatic and relative sea-level fluctuation, as well as short-term tidal, seasonal, and storm effects.
- *Coriolis Force*: influencing the movement of both air and water masses.

Autogenic controls
- *Local effects*: physical, biological, and chemical processes.
- *Post-depositional processes*: deformational, biogenic and chemogenic.
- *Progradation and aggradation*: related to sediment supply, accommodation space and accumulation.

Not only is it important to understand these controls, but it is necessary to know how they operate and at what rates, and also the rates at which sediment accumulates in different environments. Without this information it would be equally possible to interpret a small-scale sandstone–mudstone cycle, for example, as the result of sea-level change and transgression rather than due to a tidal cycle of flood-surge followed by slack water. These are clearly widely different interpretations of the same data! Some guide to the rates at which

different processes operate and different facies accumulate is given in *Fig. 15.3* (see page 261). For further information on how the controls operate see the key references listed.
Key references: Reading (1986, 1996); Walker and James (1992), Allen (1997), Leeder (2000).

Depositional environments

(*Plates* 15.33–15.63, Tables 15.2–15.13)

SEDIMENTS are presently deposited in a wide range of sedimentary environments across the surface of the Earth. Most facies from ancient sedimentary successions can be interpreted to have formed in one or other of these environments, so the study of modern analogues is very important in the interpretation of the ancient rock record. Some facies do not appear to have a modern analogue – e.g. banded iron formations. For others, we simply have not yet found a good analogue – e.g. deep-water massive sandstones.

One of two principal goals in the analysis and interpretation of sedimentary rocks in the field is to pull together all the disparate observations on sediment facies and their features, facies models and processes, facies associations, cycles and sequences, lateral trends and geometries, architectural elements and their spatial–temporal arrangement, and any information on controls and rates of sedimentation. These are then used to infer the depositional system or part of the depositional environment in which sedimentation took place. Much modern sedimentary work has been directed towards this end, and there are a number of good texts to consult (see key references).

Beyond the scope of this book and, generally, beyond the scope of field geology alone, is the further use of environmental interpretations in sequence stratigraphy, basin analysis, regional paleogeographical studies, sea level/climate change history, and plate tectonic reconstructions.

Key references: see full list at end of this chapter.

Table 15.2	Principal sedimentary environments
Main environment	**Sub-environment**
Alluvial–fluvial	Alluvial fan
	Braided river
	Meandering river
Desert–eolian	Low latitude deserts
	High latitude deserts
Lacustrine	Clastic lakes
	Carbonate lakes
	Evaporitic lakes
	Mixed lakes
Glacial	Glaciofluvial
	Glaciolacustrine
	Glaciomarine
	Periglacial
Volcaniclastic	Subaerial
	Subaqueous
	Epiclastic
Clastic coasts	Coarse-grained coasts/deltas
	River deltas
	Estuaries
	Beaches
Shallow clastic seas	Tide-dominated
	Wave-dominated
	Storm-dominated
Marine evaporites	Sabkhas
	Shallow-water evaporites
	Deep-water evaporites
Marine carbonates	Carbonate platforms
	Carbonate reefs
	Carbonate slopes
	Carbonate basins
Deep seas	Submarine fans
	Submarine slope aprons
	Submarine basins
	Contourite drifts
	Oceanic ridges
	Oceanic plateaus
	Seamounts

Alluvial–fluvial environments

15.33 Modern alluvial fan incised into Pleistocene fan succession. Evidence of neotectonic uplift is the steep vertical face at the margin of the Pleistocene succession.
Width of view 500m.
Andes Mountains, NW Argentina.

15.34 Detail of Pleistocene alluvial fan succession from the steep vertical face seen in 15.33 (above). Note poorly sorted, coarse-grained, crudely stratified deposits, typical of this environment.
Pleistocene, Andes Mountains, NW Argentina.

15.35 Fluvial red bed succession. Cliff height approx. 320m. Photograph from British Geological Survey collection, Edinburgh.
Devonian Old Red Sandstone, Old Man of Hoy, Orkney, Scotland

Desert–eolian environments

15.36 Eolian sandstone succession. Note modern, blown sand in foreground.
Width of view 10m.
Plio–Pleistocene, Kangaroo Island, S Australia.

15.37 Eolian sandstone succession with large-scale cross-stratification. This is part of the Navajo Sandstone in Zion National Park.
Section height approx. 50m.
Photo by WK Hamblin, courtesy of Paul Potter.
Triassc, Utah, USA.

15.38 Eolian sandstone succession, a productive oil-bearing reservoir sandstone in the Reconcavao Rift Basin. Width of view 10m.
Photo by Paul Potter.
Cretaceous, Bahia, Brazil.

Glacial environments

15.39 Glacial tillite – a poorly sorted pebble/boulder-rich mudstone, also associated with glacial outwash conglomerates and varved mudstones (glacial lake deposits). Pencil 12cm.
Photo by Paul Potter.
Middle Proterozoic, Sioux Ste Marie, Ontario, Canada.

15.40 Glacial outwash fluvial deposits – large-scale cross-stratified pebbly sands and sandy gravels forming the thick fill of an abandoned part of the Ohio River. Hammer 35cm.
Photo by Paul Potter.
Late Quaternary (Wisconsin), Hamilton County, Ohio, USA.

Deltaic environments

15.41 Small-scale delta system prograding into Pliocene lacustrine succession, with clear Gilbert delta topset–foreset–bottomset (T–F–B) unit. Width of view 8m. *Pliocene, S Cyprus.*

15.42 Braid-delta succession, showing marlstone–siltstone–sandstone alternation near base (pro-delta deposits), and large-scale cross-stratification in centre and upper view (Gilbert-type foresets, F, two units). Red paleosol horizons indicted by arrows. Width of view 5m. *Plio–Pleistocene, Pissouri Basin, S Cyprus.*

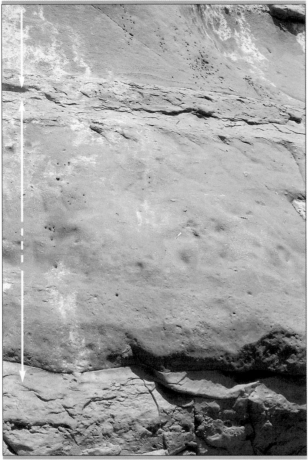

5.43 Detail of pro-delta succession in 15.42, showing reverse to normal grading (arrows) of typical hyperpycnite deposit. These hyperpycnite sands and silts were deposited on the delta slope in front of the Pissouri braid/fan delta system (15.42 and 15.44).
Width of view 60cm.
Plio–Pleistocene, Pissouri Beach, S Cyprus.

15.44 Braid-delta channel sandstones with large-scale Gilbert delta foresets (F) over bottomsets (B). Note handsome geologist for scale, pointing to foreset lamination and standing against thick, red paleosol horizon.
Pleistocene, Pissouri, S Cyprus.

Volcaniclastic environments

15.45 Subaerial volcaniclastic fall and flow succession. See other examples in Chapter 14.
Quaternary, Teide Volcano, Tenerife.

15.46 Submarine volcaniclastic turbidite succession, mainly epiclastitic deposits in forearc slope basin. See other examples in Chapter 14.
Miocene, Miura Basin, south central Japan.

Shallow-marine clastic environments

15.47 Shallow-marine foreshore to lagoonal clastic succession; note low-angle erosional scour near base (dashed line), then parallel laminated sandstones passing upwards into highly bioturbated and burrowed unit.
Width of view 1.5m.
Miocene, Sorbas Basin, SE Spain.

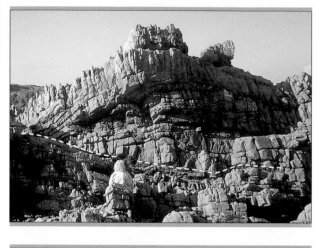

15.48 Shallow-marine tidal clastic succession, with lenticular tidal channel-fill sandstone unit cutting into thin silt–mudstone unit and more parallel-bedded sandstones. Dashed line marks base of tidal channel fill.
Triassic–Jurassic, Los Molles, west central Chile.

15.49 Shallow-marine offshore to foreshore muddy sandstone succession, with bands of carbonate concretions and intense bioturbation.
Width of view 7.5m.
Jurassic, Bridport, S England.

15.50 Deltaic mudstone–sandstone succession, with variable facies, large-scale cross-lamination, and local coal/rootlet horizons.
Width of view 10m.
Jurassic, Whitby, NE England.

Marine evaporites and carbonates

15.51 Marine evaporite succession. See other examples in Chapter 11.
Late Miocene, Sorbas Basin, SE Spain.

15.52 Carbonate knoll reef (microbial-dominated bioherm, left of centre) over gypsum surface. Knoll reef 5m wide at base. See other examples in Chapter 7.
Late Miocene, S Cyprus.

15.53 Waulsortian carbonate mud mound (Muleshoe bioherm), typical of a quiet deeper-water environment. Width of view approx. 12m. Photo by Paul Potter). *Carboniferous (Lower Mississippian), Sacramento Mountains, New Mexico, USA.*

15.54 Uplifted Quaternary reef succession surrounded by Recent coral lagoon–reef complex. Cliff height 15m. See other examples in Chapter 7. *Island of Guam, west central Pacific.*

15.55 Shallow-marine shelf-edge limestone–marl succession. See other examples in Chapter **7**.
Cretaceous, Sierra Helada, SE Spain.

15.56 Shallow-marine to lagoonal, limestone–marl–mudstone succession. See other examples in Chapter **7**.
Jurassic, Osmington Mills, S England.

Deep-marine environments

15.57 Deep-water slope to basin, well-bedded limestone succession. Whitish beds are mainly calcarenite turbidites; pinkish beds are calcilutite turbidites and hemipelagites.
Cretaceous, Umbria-Marche, Italy.

15.58 Deep-water slope to basin, chalk–chert succession. Chert bands are slightly darker, buff-coloured beds.
Width of view 6m.
Paleogene, Lefkara, S Cyprus.

15.59 Deep-water slope apron to basin, sandstone–mudstone turbidite succession. Section thickness to left of shadow approximately 120m.
Carboniferous, Andes Mountains, NW Argentina.

15.60 Deep-water turbidite sandstone–mudstone beds in mid to distal submarine-fan succession. Bed thickness variation in part caused by compensation-cycle effect. Note earnest group of turbidite specialists and petroleum geologists with Arnold Bouma (red sweater around waist). *Paleogene, S California, USA.*

15.61 Deep-water thick-bedded sandstone and pebbly sandstone turbidites of the Brushy Canyon Formation. The channel-fill body (near roadside section, approx. 5m high) is cut into thin-bedded finer-grained turbidites. Note good view of El Capitan carbonate reef as prominent rock face in the background. Photo by Paul Potter. *Permian, Culbertson County, Texas, USA.*

15.62 Deep-water turbidites and hemipelagites in basinal succession. Many of these are siliceous-rich smarls and sarls (see also Chapter **8**). Width of view in foreground 6m. *Miocene, Monterey Formation, west central California, USA.*

15.63 Deep-water thick-bedded sandstone turbidites and slump unit (S) within mid-submarine fan succession. *Paleogene, S California, USA.*

Table 15.3 Diagnostic features of alluvial–fluvial environments

Occurrence

Subaerial sediment transport pathway from upland areas to lake or marine basin; includes alluvial fan, braided, and meandering river subenvironments; all scales and types of river, all climates and parts of the world, except those with permanent snow cover.

Architectural elements/geometry

Alluvial fans: avalanche and slide mounds and wedges, shallow ephemeral channels and tadpole lobes.

Braided/meandering rivers: ribbon, clustered and belt channels, channel segments, levee ribbon mounds and crevasse splay lobes, overbank and flood-plain sheets.

Principal facies and processes

Characterized by *red bed association* – reddish or brownish-coloured conglomerates, sandstones, mudrocks, paleosols, and duricrusts; and locally coals.

Associated facies: eolian, evaporite, lacustrine, deltaic, and estuarine facies.

Depositional processes: mass gravity processes (rockfall, sliding, slumping, debris flow) especially proximally, stream-flow processes throughout, normal and flood conditions; pedogenic and coal-forming processes locally.

Characteristic features

Structures: parallel and cross-stratification at all scales, especially medium-scale cross-bedding in sandstones and indistinct stratification in conglomerates; varied erosional scours; unidirectional paleocurrents with variable dispersion; flood and sudden overbank deposits can mimic turbidites; little bioturbation, some vertebrate tracks and rootlets; irregular calcareous nodules in paleosols.

Textures and fabric: texturally immature, especially proximally; very large boulders to very fine muds; imbrication common in stream deposits.

Composition: compositionally immature, especially proximally; lithic, arkosic and locally carbonaceous sandstones common; composition reflects local source area; colour – reddish colouration typical due to oxidation of iron minerals.

Biota: general lack of fossils; where present, plants dominate, both *in situ* and fragments; fish and freshwater molluscs may occur.

Vertical Sequences

Alluvial fans: stacked coarsening-up, fining-up, and blocky mesosequences depending on nature of control.

Braided rivers: blocky and slight fining-up mesosequences.

Meandering rivers: fining-up mesosequences of cross-bedded sandstone to ripple-bedded sandstone to mudstone with paleosol, calcrete (+ coal).

Plates: 3.47–3.50, 3.121, 4.2, 4.4–4.6, 5.13, 12.9, 12.10, 13.9–13.12, 15.6, 15.13, 15.16, 15.33–15.35

Table 15.4 Diagnostic features of desert environments

Occurrence

Subaerial, low-precipitation (<10 cm/year), arid regions, generally 10–30° latitudes, or in the rain shadow of mountains and interior parts of continents; include ephemeral lakes and rivers; bordered by semi-arid environments showing many similar characteristics.

Architectural elements/geometry

Bare rock: upland areas fringed with rocky talus slopes and ephemeral alluvial fans; stony desert – pebble, rock, or silcrete pavements left as a result of deflation.

Sandy desert: sand seas, isolated/grouped seif, barchan, and other dune forms, both sheet-like and elongate geometries.

Ephemeral rivers: deep gorges through uplands, broad shallow valleys across desert plains, generally dry and poorly vegetated, rarely in flood.

Desert lakes: typically broad, shallow depressions, episodically filled or partially filled with water from internal drainage, regions of evaporite/sabkha facies.

Principal facies and processes

As for alluvial/fluvial successions, desert sediments are characterized by the **red bed association** – alluvial fan/fluvial conglomerates, (pebbly) sandstones and mudstones; eolian sandstones with isolated pebbles; various evaporites and duricrusts.

Associated facies: lacustrine mudstones/siltstones; wind-blown silts (loess).

Depositional processes: rockfall, debris flow and stream flow, eolian processes, evaporitic processes.

Characteristic features

Structures: generally chaotic alluvial/fluvial facies; sandstones show large-scale and complex cross-bedding, with foresets dipping up to 35°, parallel lamination, reactivation surfaces.

Textures and fabric: poorly sorted immature fluvial facies, interbedded with well-sorted, well-rounded sands, and wind-polished wind-faceted pebbles; sand grains may have dull frosted suface texture.

Composition: immature to sub-mature; composition reflects local source area; common evaporite minerals; colour – reddish colouration typical due to oxidation of iron minerals.

Biota: rare, some vertebrate bones and tracks, rootlets and plant debris in fluvial facies.

Vertical Sequences

Irregular interbedding of different desert facies; some cyclicity may be apparent in evaporite deposits from longer-lived lakes or marine incursion events.

Plates: 3.13–3.31, 3.44, 3.45, 3.121, 6.18, 11.8–11.15, 13.4–13.8, 15.36–15.38

Table 15.5 Diagnostic features of lacustrine environments

Occurrence

Local depocentres formed on continental landmasses; occur in all sizes, shapes, and climates; they show a huge range of scale and features so that generalizations are difficult. Lakes can be <100m^2 in area to >82,000km^2 (e.g. Lake Superior), very deep set in tectonic rifts (e.g. Lake Baikal reaches 1620m), to very shallow and ephemeral desert lakes (e.g. Lake Eyre). Larger lakes present a natural laboratory of other environments – deltaic, coastal, slope, fan and deep-water.

Architectural elements/geometry

No single style, element, or geometry characterizes lacustrine environments; all elements from coastal, through slope, to deep water can occur. Lake basin facies tend to show sheet-like geometry.

Principal facies and processes

Diverse facies, including breccias, conglomerates, sandstones, mudrocks and oil shales; limestones, marls, and cherts; coals, evaporites, and volcaniclastic sediments.

Associated facies: alluvial fan, fluvial, eolian, fan-delta, and deltaic facies are all common.

Depositional processes: waves, surface and storm currents in shallow water; hyperpycnal flows, turbidity currents, and mass-gravity processes in slope and deep basinal areas; biochemical and chemical evaporation/precipitation processes also common; volcaniclastic and glacio–lacustrine processes where applicable. Strong salinity and climate control on sedimentation.

Characteristic features

Structures: shallow-water structures, including wave ripples, desiccation cracks, and rainspots; microbial carbonate lamination and stromatolites; deeper-water slump features, graded beds, and turbidite structures; rhythmic layering and fine varve-type lamination in quiet basinal waters.

Textures and fabric: coarse-grained and poorly sorted lake margin facies rapidly pass basinwards into fine-grained facies; extensive mud and micrite sediments are common of many lakes.

Composition: each of the principal facies groups can be represented in lakes, reflected in siliciclastic, carbonate, chemogenic, and volcaniclastic compositions; finer-grained sediments can be rich in organic matter, including freshwater algae and coals.

Biota: rich in non-marine fossils including plants, bivalves, gastropods, vertebrate bones, and tracks

Vertical Sequences

These reflect changes in climate and tectonics, which in turn influence sediment supply and water level; shallowing-upward sequences are common, capped by coals, evaporites, fluvial, or paleosol facies.

Plates: 3.20, 3.28, 3.30, 6.17, 11.9–11.15, 14.17, 14.18, 15.41

15

Table 15.6   Diagnostic features of glacial environments

Occurrence

Characteristic of high latitude and high mountainous regions – ice covers approx. 10% of today's world; more extensive coverage (approx. 30% ice cover) and influence during past glacial periods. Range from proximal erosive features to more distal depositional facies on land; broad areas of periglacial conditions and glacial outwash; and extensive glaciolacustrine and glaciomarine influence.

Architectural elements/geometry

Irregular to regular erosive scours, lineations, and polished surfaces, deeply incised U-shaped valleys and fjords, over-deepened lakes and shelf basins; irregular depositional masses of lodgement and deformational till, hummocks, eskers, and subglacial channel bodies; more regular proglacial, periglacial, and glaciomarine bodies.

Principal facies and processes

Diverse facies suite characterized by distinctive polymict muddy conglomerates (diamictites or tillites), rhythmically varved silty mudstones, and mudstones with dispersed clasts (dropstones). Other facies include various conglomerates, sandstones and mudstones, some with glacial deformation and cryoturbation.

Associated facies: any normal continental, lacustrine, or marine facies deposited in periglacial regions influenced by alternating glacial advance and retreat. Ice-rafted debris can occur in normal marine facies far removed from the region of iceberg calving.

Depositional processes: mass balance between ablation and accumulation determines whether glaciers advance, retreat, or remain stationary, and hence control deposition from glaciers; other processes include meltwater and subglacial streams, proglacial density currents and debris flows, and glacial deformation by loading, bulldozing, and freeze–thaw mechanisms.

Characteristic features

Structures: tillites are generally structureless, glaciolacustrine mudstones may show seasonal varves (lamination) with or without dropstones, and glaciomarine deposits typically showed randomly dispersed dropstones; typical fluvial features occur in glaciofluvial facies (Table 15.4), and the same for glaciolacustrine and glaciomarine facies; striated glacial pavements and boulders, soft-sediment scours, and a range of glacial deformation structures are common; frost action creates a variety of deformation structures and fabric reorganization known as cryoturbation.

Textures and fabric: from poorly sorted muddy gravels, pebbly (and boulder) mudstone to finely varved mudstone; random to regular fabric depending on facies type.

Composition: often polymict and very varied composition; far-travelled glacial erratics and ice-rafted debris may have an exotic composition.

Biota: generally devoid of fossils.

Vertical Sequences

Random sequences of facies are most common.

Plates: 3.10–3.12, 4.14, 6.7, 6.8, 15.39, 15.40

Table 15.7   Diagnostic features of volcaniclastic environments

Occurrence

In association with all present and past areas of volcanic activity, particularly active plate-tectonic settings. May be subaerial, subglacial, sublacustrine, and submarine. Epiclastic sediments may be reworked far from the original volcanic source, and volcanic ash can be dissipated around the globe.

Architectural elements/geometry

Irregular and chaotic deposits close to source; sheet-like pyroclastic fall drapes over topography; complete or partial concentration of pyroclastic surge deposits along topographic depressions; normal diverse range of elements and geometries associated with epiclastic reworking – from fluvial, through coastal, to deep-marine environments.

Principal facies and processes

Characterized by chaotic and fragmented facies proximally, and well-bedded/structured facies distally. Autoclastites, pyroclastic fall, flow and surge facies, hydroclastites and epiclastites; massive chaotic fall agglomerates and breccias, intermixed lava-wet sediment peperites, volcaniclastic sandstones with flow banding and accretionary lapilli, and thin-bedded, fine-grained, laterally extensive tuffs are all diagnostic facies.

Associated facies: depends entirely on location of volcaniclastic source with respect to other environments; important to distinguish subaerial from subaqueous facies.

Depositional processes: autobrecciation and hydroclastic fragmentation, air-fall and water-fall processes, pyroclastic flows and surges, and reworking by normal surface processes – fluvial, lacustrine, shallow-marine waves and currents, and downslope re-sedimentation.

Characteristic features

Structures: primary deposits are commonly chaotic and structureless, also graded; some beds show lamination, cross-lamination, streaked-out fiamme, and flow banding; large-scale wavy or antidune cross-stratification occurs in surge facies; secondary deposits (epiclastites) show full range of sedimentary structures.

Textures and fabric: very diverse textures, from very coarse and poorly sorted to very fine and well sorted (proximal to distal trends); welding in hot pyroclastic flows (ignimbrites) and matrix support in lahars.

Composition: wholly volcaniclastic composition including primary crystals (especially feldspars), quenched glass, and acid to basic lithic clasts; chemical instability leads to rapid alteration to diagenetic clay minerals.

Biota: fossils can occur, especially in pyroclastic fall deposits and in epiclastites, but are generally rare, except in some more distal water-lain facies.

Vertical Sequences

No particular facies sequences apparent.

Plates: 3.7, 3.25, 3.26, 3.55, 4.13, 14.1–14.29, 15.19, 15.45, 15.46

Table 15.8   Diagnostic features of deltaic environments

Occurrence

Major depocentres where rivers flow into lakes or the sea, that occur on variety of scales from <1 km^2 to >100,000 km^2 in area and with up to at least a 15km sediment pile at its thickest. Coarse-grained deltas are fed directly by alluvial fans or braided gravel rivers. Classic river deltas have significant mixed-grade river input, and in some the sediments are reworked by wave or tidal processes.

Architectural elements/geometry

Delta-top elements include distributary channels and levees, lakes, swamps, and marsh-lands, river mouth and distal bars, and interdistributary bays. Delta-front elements include the prodelta slope, slope channels and levees, and slide–slump masses. These grade downslope into delta-bottom units with largely sheetlike and lenticular geometry.

Principal facies and processes

River deltas are characterized by sandstones, siltstones, and mudstones (plus inter-mediate siliciclastic facies) of many different facies. Coals, paleosols, and ironstones are common in delta-top settings. Coarse-grained deltas also have a dominance of conglom-erates and breccias.

Associated facies: fluvial (and alluvial fan) facies, shallow marine sediments, and deeper basinal facies (for large deltas).

Depositional processes: these range from river-dominated processes(channel flow, spill-over and overbank settling), through marine currents, waves,and tides, to downslope processes on the delta front; channel collapse and large-scale delta-front slumping are common where sedimentation rates are very high; coal-forming processes also occur on the delta top.

Characteristic features

Structures: current-flow primary structures very common, especially cross stratification, together with flaser and wavy bedding in the finer-grained facies. Slump structures and bioturbation also common in parts.

Textures and fabric: full range of grain size, sorting and fabric.

Composition: siliciclastic-dominated facies; may be carbonaceous-rich and with distinct coal seams; some ironstones and iron-rich concretions (especially siderite).

Biota: Marine fossils common in more distal facies, non-marine and intermediate fossils in more proximal facies. Plant debris very common, as is bioturbation and burrowing.

Vertical Sequences

Delta progradation produces medium to large-scale coarsening-up facies sequences, capped by seat earths and coals. Marine transgression and delta lobe abandonment can lead to fining-up facies sequences, capped with marine mudstones and, in some cases, limestones. Several sequence variations occur locally due to channel fill, channel switch-ing and abandonment, interdistributary bay fill, and others.

Plates: 3.24, 3.57, 5.15, 10.1–10.6, 15.41–15.44, 15.50

Table 15.9 Diagnostic features of shallow-marine clastic environments

Occurrence

From the coastline to outer shelf and epeiric seas, these environments include beaches, barrier islands, estuaries, lagoons, tidal flats, shoreface, and open shelves, where siliciclastic sediment input and reworking dominates.

Architectural elements/geometry

Diverse suite of elements depending on particular subenvironment – sheets, wedges and lenses, ribbons, linear sands, and channel bodies; zones of non-deposition and winnowing leaving hiatuses and condensed sections.

Principal facies and processes

Thin, well-rounded conglomerates, sandstones (often mature/super mature), muddy and calcareous (shell-rich) sandstones, siltstones and mudstones. Glauconitic sandstones, ironstones, and phosphorite concentrations occur locally.

Associated facies: interdigitate with carbonate/evaporite facies especially at low latitudes, glaciomarine facies at high latitudes; and with deep-water slope sediments and continental facies during transgressive–regressive episodes.

Depositional processes: waves, shallow marine and tidal currents, and storm events are the dominant processes.

Characteristic features

Structures: parallel and cross-stratification at all scales are typical of the varied current processes in action, including herringbone cross-lamination, wave ripples, flaser and lenticular lamination, hummocky cross-stratification and swaley cross-stratification; truncation surfaces, mud drapes and thin storm-graded beds common; abundant bioturbation and trace fossils, carbonate and pyrite concretions.

Textures and fabric: from very fine-grained to coarse pebbly sands, some with high textural maturity; evidence of winnowing and reworking quite common.

Composition: sandstones may be mature to supermature, rich in shallow-marine biogenic material, glauconitic, micaceous or iron-rich; more or less calcareous depending on clastic input vs. biogenic influence.

Biota: marine fossils and trace fossils common, both showing shallow-water affinities, with diversity dependent on substrate, salinity, energy level, etc.

Vertical Sequences

Facies sequences much affected by sea-level change, tending to coarsen-up with sea-level fall and fine-up with sea-level rise. Many other more complex facies sequences and associations also occur.

Plates: 3.18, 3.22–3.35, 3.37, 3.38, 3.40–3.43, 3.46, 3.98–3.100, 4.7, 5.7, 5.10–5.12, 5.14, 5.16, 6.12, 6.13, 6.16, 12.11–12.16, 15.11, 15.47–15.50

Table 15.10 Diagnostic features of marine evaporite environments

Occurrence

Marine evaporites are most typical of arid regions of the world, especially along the coastline and offshore of desert regions, and beneath or surrounding enclosed and semi-enclosed seas. They occur where evaporation exceeds precipitation and runoff.

Architectural elements/geometry

Generally thin sheet-like elements, in some cases elongated parallel to shore.

Principal facies and processes

Range of evaporite facies, principally sulfates (gypsum and anhydrite), and halides (halite, polyhalite). More exotic evaporites characterize continental environments.

Associated facies: carbonates, including microbial laminites, oolites, and dolomites; and varied siliciclastic facies, often closely interbedded with evaporite facies.

Depositional processes: evaporative chemogenic processes dominant, together with shallow marine currents, tides and waves.

Characteristic features

Structures: nodular, massive, and laminated – evidence of either *in situ* precipitation or reworking by marine currents. Chicken-wire and tepee structures, microbial lamination, dessication cracks, raindrops, tracks and trails. Crystal pseudomorphs in mudstone common – eg halite cubes.

Textures and fabric: crystalline texture and crystal-welded fabric; also crystal grains and clasts.

Composition: Evaporite minerals (sulfates and halides dominant), together with associated carbonates and siliciclastic components.

Biota: Fossils very rare in evaporites but salt-tolerant species, including microbial and algal lamination and fossils occur in associated facies, together with limited surface tracks and trails.

Post depositional changes: Note that evaporites are very subject to post-depositional dissolution, transport, and re-precipitation. Many pre-Cenozoic evaporites, therefore, will show secondary rather than primary features.

Vertical Sequences

No particular diagnostic sequences, but cycles of alternation between evaporite and non-evaporite facies are typical.

Plates: 11.1–11.7, 15.51, 15.52

Table 15.11 Diagnostic features of shallow-marine carbonate environments

Occurrence

From the coastline to outer shelf and epeiric seas, in generally low latitudes without significant input of siliciclastic sediment. Environments include carbonate-rich beaches, barrier islands, banks and shoals, lagoons, tidal flats, shoreface, open shelves, and reefs of all different kinds.

Architectural elements/geometry

Diverse suite of elements depending on particular subenvironment – sheets, mounds and lenses, ribbons, linear carbonate sands and channel bodies; reef build-ups and talus wedges; zones of non-deposition and winnowing, leaving hiatuses and condensed sections.

Principal facies and processes

Many different limestone facies, as well as primary and secondary dolomites; especially oosparites, oomicrites, biosparites, biomicrites, and biolithites.

Associated facies: siliciclastic and mixed clastic–carbonate facies, evaporites and cherts are common associates.

Depositional processes: biological and biochemical processes dominate in the origin of sediment grains and bioclasts; these are then variously affected by waves, storms, tides, and other marine currents.

Characteristic features

Many features are similar to those of shallow-marine siliciclastic environments, whereas others are very specific to carbonate facies.

Structures: Parallel and cross-stratification at all scales are typical of the varied current processes in action in shallow marine settings. A variety of reef and associated features also occur – *in situ* framework and fragmented talus, stromatolites and microbial lamination, fenestrae, stromatactis, and stylolites.

Textures and fabric: Varied, from very fine to very coarse-grained, from cemented boundstone to coarse, fragmented calcirudite; pebbly sands, some with high textural maturity; evidence of winnowing and reworking quite common.

Composition: Carbonate dominated – micrite, sparite, ooids, pisoids, oncoids, and bioclastic material. Variable admixture of terrigenous impurities, evaporites and siliceous cementation. More rarely phosphatic or carbonaceous.

Biota: Marine fossils and trace fossils common, both showing shallow-water affinities, with diversity dependent on substrate, salinity, energy level, etc.

Vertical Sequences

Facies sequences much affected by sea-level change.

Plates: 7.1, 7.2, 7.5–7.21, 7.26, 8.2, 8.3, 8.5, 8.6, 15.5, 15.12, 15.13, 15.17, 15.53–15.56

Table 15.12 Diagnostic features of deep-marine carbonate environments

Occurrence

Very widespread across the slope aprons surrounding carbonate shelves, banks and reefs, and throughout the deep sea. Carbonate is not generally preserved at very great depths (except when rapidly buried) due to its ready dissolution in seawater below the carbonate compensation depth (CCD). The level of the CCD varies through time and space, currently being around 4–5km in the Atlantic and 3–4km in the Pacific.

Architectural elements/geometry

Slope-apron elements include channels and scours, irregular slide–slump and debrite mounds, reef talus wedges, contourite mounds, turbidite lobes and sheets. Basinal elements are mostly sheet-like or mounded in geometry.

Principal facies and processes

Limestone facies include coarse-grained to fine-grained resedimented calcirudites, calcarenites and calcilutites (rockfall, slide, slump, debrite, and turbidite facies); and calcilutite to calcarenite contourite, hemipelagite and pelagite facies.

Associated facies: cherts, phosphorites, ferromanganese nodules and crusts, and a variety of deep-water siliciclastic or volcaniclastic facies.

Depositional processes: pelagic, hemipelagic, and resedimentation processes are dominant; bottom current and chemogenic processes are locally important.

Characteristic features

Structures: resedimented facies are structureless/chaotic, or have slump folds, debrite and turbidite sequences of structures (mostly partial sequences), grading and water-escape structures; contourite facies mostly have very subtle contourite features and much bioturbation, as illustrated in the contourite facies/sequence model, and are often difficult to distinguish from hemipelagites; pelagites and hemipelagites can be nodular and heavily bioturbated, or laminated; hardgrounds, ferromanganese encrustation, and hiatuses are common.

Textures and fabric: varied, from very fine to very coarse-grained; calcilutite to calcirudite; mudstone, wackestone and floatstone common, with some packstone and rudstone.

Composition: carbonate dominated – micrite, sparite and bioclastic material; planktonic microfossils (in pelagites) and resedimented shallow-water bioclasts and intraclasts (in calciturbidites); siliceous microfossils and diagenetic silica, ferromanganese and phosphorite minerals occur locally; variable admixture of siliciclastic material; organic carbon locally significant.

Biota: fossils and trace fossils variably common, especially microfossils in finer-grained and pelagic facies; derived shallow-water fossils in resedimented facies.

Vertical Sequences

No very typical sequence of resedimented carbonate facies, more a random interbedding. Small-scale cyclicity common in pelagic, hemipelagic, and contourite successions.

Plates: 3.14, 7.23–7.25, 15.1, 15.2, 15.58, 15.62

Table 15.13 Diagnostic features of deep-marine clastic environments

Occurrence

Very widespread across slope aprons, submarine fans, contourite drifts, and basin plains. Occurring in oceanic basins and marginal seas, with parallels in inland seas and large lakes.

Architectural elements/geometry

Slope-apron elements include channels and scours, irregular slide–slump and debrite mounds, contourite mounds and other drifts, turbidite levees, lobes and sheets. Submarine-fan elements as for slope aprons, although if contourites occur they form combination fan-drift elements. Basinal elements are mostly sheet-like or mounded in geometry, but with some small channel, levee, and lobe geometrics.

Principal facies and processes

Rockfall, slide, slump, debrite, and turbidite facies, interbedded with contourite, hemipelagite, and pelagite facies. Below the CCD, pelagites are typically very fine-grained red and brown clays. Glaciomarine hemipelagites are typical of high latitudes.

Associated facies: cherts and deep-water carbonate facies, ferromanganese nodules and crusts, and a variety of deep-water volcaniclastic and glaciomarine facies.

Depositional processes: pelagic, hemipelagic, and resedimentation processes are dominant; bottom current and chemogenic processes are locally important.

Characteristic features

Structures: resedimented facies are structureless/chaotic, or have slump folds, debrite and turbidite sequences of structures (mostly partial sequences), grading and water-escape structures; contourite facies mostly have very subtle contourite features and much bioturbation, as illustrated in the contourite facies/sequence model, and are often difficult to distinguish from hemipelagites; pelagites and hemipelagites are generally heavily bioturbated, more rarely laminated; winnowed horizons, ferromanganese encrustation and hiatuses are less common than with deep-marine carbonates.

Textures and fabric: very varied, from coarse-grained conglomerates and breccias to very fine-grained mudstones.

Composition: siliciclastic dominated, immature to mature in composition; variable admixture of carbonate, typically as microfossils and resedimented shallow-water bioclasts, and as siliceous microfossils; volcaniclastic material occurs locally and may be dominant; organic carbon locally abundant (black shale facies).

Biota: Fossils and trace fossils variably very rare to common, especially microfossils in finer-grained and pelagic facies; derived shallow-water fossils and carbonaceous debris in re-sedimented facies.

Vertical Sequences

A variety of small and medium-scale facies sequences are recognized in deep-marine successions – fining-up, coarsening-up, symmetrical, blocky – and indicate different architectural elements and process sequences. Irregular and random interbedding of facies also occurs. Small-scale cyclicity common in pelagic, hemipelagic, and contourite successions.

Plates: 3.5–3.7, 3.16–3.23, 3.36–3.39, 3.51–3.74, 3.82–3.90, 3.92–3.95, 4.1, 4.3, 4.9, 4.10, 4.12, 4.15, 4.16, 5.1–5.63, 5.17, 5.18, 6.1–6.3, 6.5, 6.6, 6.9, 6.10, 6.14, 6.20–6.22, 7.25, 8.4, 8.9–8.12, 12.19–12.21, 15.3, 15.4, 15.7–15.10, 15.14, 15.15, 15.20, 15.57–15.63

REFERENCES AND KEY TEXTS

Adams AE, MacKenzie WS, Guilford C, (1984) *Atlas of Sedimentary Rocks Under the Microscope*, Longman.

Adams AE, MacKenzie WS, (1994) *A Colour Atlas of Carbonate Sediments and Rocks Under the Microscope*, Manson Publishing, London.

Allen JRL, (1982) *Sedimentary Structures, Vols 1 and 2*, Elsevier, Amsterdam.

Allen JRL, (1985) *Principles of Physical Sedimentology*, Unwin-Hyman, London.

Allen PA, Allen JRL, (1990) *Basin Analysis: Principles and Applications*, Blackwell Science, Oxford.

Allen PA, (1997) *Introduction to Earth Surface Processes*, Blackwell Science, Oxford.

Bathurst RGC, (1975) *Carbonate Sediments and their Diagenesis*, Elsevier, Amsterdam.

Blatt H, (1992) *Sedimentary Petrology*, Freeman, San Francisco.

Blatt H, Middleton GV, Murray R, (1980) *Origin of Sedimentary Rocks*, Prentice Hall, New Jersey.

Boggs S, (1992) *Petrology of Sedimentary Rocks*, Prentice Hall, New Jersey.

Boggs S, (1995) *Principles of Sedimentology and Stratigraphy*, Prentice Hall, New Jersey.

Bouma AH, (1969) *Methods for the Study of Sedimentary Structures*, Wiley-Interscience, New York.

Brenner RL, McHargue T, (1988) *Integrative Stratigraphy*, Prentice-Hall, New Jersey.

Bromley RG, (1990) *Trace Fossils: Biology and Taphonomy*, Unwin-Hyman, London.

Busby CJ, Ingersoll RV (eds) (1995) *Tectonics of Sedimentary Basins*, Blackwell Science, Oxford.

Carver RE (ed) (1971) *Procedures in Sedimentary Petrology*, Wiley, New York.

Cas RAF, Wright JV, (1987) *Volcanic Successions: Modern and Ancient*, Allen & Unwin, London.

Chamley H, (1989) *Clay Sedimentology*, Springer-Verlag, Berlin.

Collinson JD, Thompson DB, (1989) *Sedimentary Structures*, Allen & Unwin, London.

De Boer PL, Smith DG (eds) (1994) *Orbital Forcing and Cyclic Sequences*, International Association of Sedimentologists Special Publication, 19.

Einsele G, (1992) *Sedimentary Basins*, Springer-Verlag, Berlin.

Emery D, Myers KJ, (1996) *Sequence Stratigraphy*, Blackwell Science, Oxford.

Fisher RV, Schmincke HU, (1984) *Pyroclastic Rocks*, Springer-Verlag, Berlin.

Flugel E, (1982) *Microfacies Analysis of Limestones*, Springer-Verlag, Berlin.

Folk RL, (1974) *Petrology of Sedimentary Rocks*, Hemphill Publishing, Austin, Texas.

Frey RW (ed) (1975) *The Study of Trace Fossils*, Springer-Verlag, New York.

Friedman GM, Sanders JE, Kopaska-Merkel DC, (1992) *Principles of Sedimentary Deposits*, Macmillan, New York.

Fritz WJ, Moore JN, (1988) *Basics of Physical Stratigraphy and Sedimentology*, John Wiley, New York.

Goldring R, (1991) *Fossils in the Field*, Longman, Essex.

Greensmith JT, (1988) *Petrology of the Sedimentary Rocks*, Allen & Unwin, London.

Hallam A, (1981) *Facies Interpretation and the Stratigraphic Record*, Freeman, Oxford.

Hedberg HD (ed.) (1976) *International Stratigraphic Guide*, Wiley InterScience.

Holland CH, et al. (1978) *A Guide to Stratigraphical Procedure*, Geological Society London Special Report, 11.

Leeder MR, (1982) *Sedimentology*, Allen & Unwin, London.

Leeder MR, (1999) *Sedimentology and Sedimentary Basins: From Turbulence to Tectonics*, Blackwell Science, Oxford.

Lewis DW, (1984) *Practical Sedimentology*, Hutchinson Ross, Stroudsburg.

Lewis DW, McConchie, D. (1994) *Analytical Sedimentology*, Chapman & Hall, London.

Lindholm RC, (1987) *A Practical Approach to Sedimentology*, Unwin Hyman, London.

Matthews RK, (1984) *Dynamic Stratigraphy*, Prentice Hall, New Jersey.

McClay K, (1987) *The Mapping of Geological Structures*, John Wiley & Sons, Chichester.

MacKenzie WS, Adams AE, (1992) *A Colour Atlas of Rocks and Minerals in Thin Section*, Manson Publishing, London.

Miall AD, (1990) *Principles of Sedimentary Basin Analysis*, Springer-Verlag, New York.

Miall AD, (1997) *The Geology of Stratigraphic Sequences*, Springer-Verlag, New York.

Muller G, (1967) *Methods in Sedimentary Petrology*, Schweizerbart, Stuttgart.

Nichols G, (1999) *Sedimentology and Stratigraphy,* Blackwell Science, Oxford.

Pettijohn FJ, (1975) *Sedimentary Rocks*, Harper & Row, New York.

Pettijohn FJ, Potter PE, (1964) *Atlas and Glossary of Primary Sedimentary Structures,* Springer-Verlag, Berlin.

Pettijohn FJ, Potter PE, Siever R, (1987) *Sand and Sandstone*, Springer-Verlag, Berlin.

Pickering KT, Hiscott RN, Hein FJ (1989) *Deep-Marine Clastic Environments: Clastic Sedimentation and Tectonics*, Unwin Hyman, London.

Posamentier HW, *et al.* (eds) (1993) *Sequence Stratigraphy and Facies Associations*, International Association of Sedimentologists Special Publication, 18, Blackwell Science, Oxford.

Potter PE, Maynard JB, Pryor WA, (1980) *Sedimentology of Shale*, Springer-Verlag, Berlin.

Potter PE, Pettijohn FJ, (1977) *Palaeocurrents and Basin Analysis*, Springer-Verlag, Berlin.

Pye K, (1994) *Sediment Transport and Depositional Processes*, Blackwell Science, Oxford.

Reading HG (ed) (1996) *Sedimentary Environments and Facies*, 3rd edn. Blackwell Science, Oxford.

Reineck HE, Singh IB, (1980) *Depositional Sedimentary Environments*, Springer-Verlag, Berlin.

Retallack G, (1989) *Soils of the Past: An Introduction to Palaeopedology*, Harper Collins Academic, New York.

Salvador A (ed) (1994) *International Stratigraphic Guide*, 2nd edn. IUGS and Geological Society of America

Selley RC, (2000) *Applied Sedimentology*, Academic Press, San Diego.

Scoffin TP, (1987) *An Introduction to Carbonate Sediments and Rocks*, Blackie & Son, Glasgow.

Scholle PA, Bebout DG, Moore CH (eds) (1983) *Carbonate Depositional Environments*, American Association of Petroleum Geologists Memoir, 33.

Scholle PA, Spearings D (eds) (1982) *Sandstone Depositional Environments*, American Association of Petroleum Geologists Memoir, 31.

Selley RC, (1996) *Ancient Sedimentary Environments*, Chapman & Hall, London.

Stach E, *et al.* (eds) (1982) *Stach's Textbook of Coal Petrology*, Lubrecht & Cramer (3rd Ed.)

Stow DAV, (2005) *Encyclopedia of the Oceans*, Oxford University Press, Oxford.

Summerhayes CP, Thorpe SA (eds) (1996) *Oceanography: An Illustrated Guide*, Manson Publishing, London.

Tucker ME (ed) (1988) *Techniques in Sedimentology*, Blackwell Science, Oxford.

Tucker ME, (1991) *Sedimentary Petrology: An Introduction to the Origin of Sedimentary Rocks*, Blackwell Science, Oxford.

Tucker ME, (2003) *Sedimentary Rocks in the Field*, 3rd edn. John Wiley & Sons, New York.

Tucker ME, Wright P, (1990) *Carbonate Sedimentology and Diagenesis*, Blackwell Science, Oxford.

Van Wagoner JC, *et al.* (1990) *Siliciclastic Sequence Stratigraphy in Well Logs, Cores and Outcrops*, American Association of Petroleum Geologists Methods in Exploration Series, 7

Walker RG, (1984) *Facies Models*, Geoscience Canada, Toronto.

Walker RG, James NP (eds) (1992) *Facies Models – Response to Sea Level Changes,* Geoscience Canada.

Ward C, (1984) *Coal Geology and Coal Technology*, Blackwell Science, Oxford.

Warren JK, (1989) *Evaporite Sedimentology*, Prentice Hall, New Jersey.

Weaver CE, (1989) *Clays, Muds and Shales*, Elsevier, Amsterdam.

Weimer P, Posamentier HW (eds) (1993) *Siliciclastic Sequence Stratigraphy: Recent Developments and Applications,* American Association of Petroleum Geologists Memoir, 58.

Wilgus CK, *et al.* (1988) *Sea Level Changes – An Integrated Approach*, Society of Economic Paleontologists and Mineralogists Special Publication, 42.

Wilson JL, (1975) *Carbonate Facies in Geologic History*, Springer-Verlag, Berlin.

Young TP, Taylor WEG, (1989) *Phanerozoic Ironstones*, Geological Society of London Special Publication, 46.

Metric–imperial conversions

Length

1 μm (micron)	=	10^{-4} cm or 10^{-6} m
1 cm	=	0.39 inches
1 m	=	3.28 feet
1 km	=	0.62 miles
1 inch	=	2.54 cm
1 foot	=	0.305 m
1 mile	=	1.61 km

Area

1 cm^2	=	0.155 square inches
1 m^2	=	10.76 square feet
1 km^2	=	0.386 square miles

Volume

1 cm^3	=	0.061 cubic inches
1 m^3	=	35.314 cubic feet

Weight

1 g	=	0.035 ounces
1 kg	=	2.205 pounds
1 tonne	=	1.1 ton (short tons)

INDEX

Page numbers in **bold** type refer to plate captions; page numbers in *italic* type refer to figures and tables.

STRATIGRAPHIC TIMESCALE

EON	ERA	PERIOD		EPOCH / AGE		Ma 248.2	AGE	Duration
PHANEROZOIC	PALEOZOIC	PERMIAN	L	Tatarian		252.1	3.9	7.8
				Ufimian – Kazanian		256.0	3.9	
			E	Kungurian		260.0	4.0	
				Artinskian		269.0	9.0	34.0
				Sakmarian		282.0	13.0	
				Asselian		290.0	8.0	
		CARBONIFEROUS	PENNSYLVANIAN (L)	Gzelian		296.5	6.5	33.0
				Kasimovian		303.0	6.5	
				Moscovian		311.0	8.0	
				Bashkirian		323.0	12.0	
			MISSISSIPPIAN (E)	Serpukhovian		327.0	4.0	31.0
				Visean		342.0	15.0	
				Tournaisian		354.0	12.0	
		DEVONIAN	L	Famennian		364.0	10.0	16.0
				Frasnian		370.0	6.0	
			M	Givetian		380.0	10.0	21.0
				Eifelian		391.0	11.0	
			E	Emsian		400.0	9.0	26.0
				Praghian		412.0	12.0	
				Lochkovian		417.0	5.0	
		SILURIAN	L	Pridoli		419.0	2.0	6.0
				Ludlow		423.0	4.0	
				Wenlock		428.0	5.0	
			E	Llandovery		443.0	15.0	20.0
		ORDOVICIAN	L	Ashgill		449.0	6.0	15.0
				Caradoc		458.0	9.0	
			M	Llandeilo		464.0	6.0	12.0
				Llanvirn		470.0	6.0	
			E	Arenig		485.0	15.0	25.0
				Tremadoc		495.0	10.0	
		CAMBRIAN	L	Merioneth		505.0	10.0	10.0
			M	St. David's		518.0	13.0	13.0
			E	Caerfai	Lenian	524.0	6.0	27.0
					Atdabanian	530.0	6.0	
					Tommotian	534.0	4.0	
					Nemakitian – Daldynian	545.0	11.0	

EON	ERA	PERIOD	EPOCH / AGE		Ma	AGE Duration		Eustatic curve
					65.0	6.3		
			Maastrichtian		71.3			250 15C 50 0 M
			Campanian	L		12.2	33.9	
					83.5			
			Santonian		85.8	2.3		
			Coniacian		89.0	3.2		
			Turonian		93.5	4.5		
			Cenomanian		98.9	5.4		
		CRETACEOUS	Albian		112.2	13.3		
			Aptian	E	121.0	8.8	43.1	
			Barremian		127.0	6.0		
			Hauterivian		132.0	5.0		
			Valanginian		136.5	4.5		
			Ryazanian / Berriasian		142.0	5.4		Long term
P H A N E R O Z O I C	M E S O Z O I C		Portlandian / Volgian / Tithonian	L	144.0 / 145.6 / 150.7	8.8	17.4	
			Kimmeridgian		154.1	3.4		
			Oxfordian		159.4	5.3		
		JURASSIC	Callovian	M	164.4	5.0		
			Bathonian		169.2	4.8	20.7	
			Bajocian		176.5	7.3		
			Aalenian		180.1	3.6		
			Toarcian		189.6	9.5		
			Pliensbachian	E	195.3	5.7	25.6	
			Sinemurian		201.9	6.6		
			Hettangian		205.7	3.8		
			Rhaetian		209.6	3.9		
			Norian	L	220.7	11.1	21.7	
		TRIASSIC	Carnian		227.4	6.7		
			Ladinian		234.3	6.9		
			Anisian	M	241.7	7.4	14.3	
			Olenekian		244.8	3.1	6.5	
			Induan	E		3.4		

STRATIGRAPHIC TIMESCALE

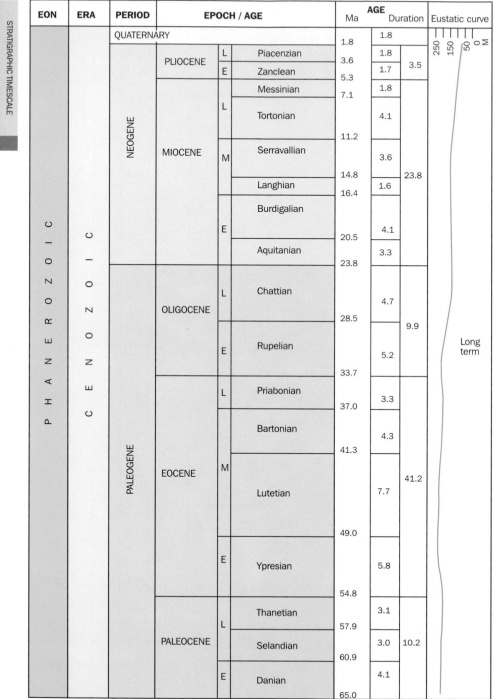

EON	ERA	PERIOD	EPOCH / AGE			Ma	AGE 1.8	Duration	Eustatic curve
		QUATERNARY				1.8	1.8		250 150 50 0 M
			PLIOCENE	L	Piacenzian	3.6	1.8	3.5	
				E	Zanclean	5.3	1.7		
			MIOCENE	L	Messinian	7.1	1.8	23.8	
					Tortonian	11.2	4.1		
		NEOGENE		M	Serravallian	14.8	3.6		
					Langhian	16.4	1.6		
				E	Burdigalian	20.5	4.1		
P H A N E R O Z O I C	C E N O Z O I C				Aquitanian	23.8	3.3		
			OLIGOCENE	L	Chattian	28.5	4.7	9.9	Long term
				E	Rupelian	33.7	5.2		
		PALEOGENE	EOCENE	L	Priabonian	37.0	3.3	41.2	
					Bartonian	41.3	4.3		
				M	Lutetian	49.0	7.7		
				E	Ypresian	54.8	5.8		
			PALEOCENE	L	Thanetian	57.9	3.1	10.2	
					Selandian	60.9	3.0		
				E	Danian	65.0	4.1		

lithotypes

- conglomerate (cgl)
- sandstone (sst)
- siltstone (slt)
- mudstone (md) claystone (cly)
- limestone (lst)
- dolomite (dol)
- chert (cht)
- coal (cl)
- halite (hal)
- gypsum, anhydite (gyp, anhy)
- volcaniclastite
- closely interbedded lithotypes; width of ornament indicates proportion of each

qualifiers

pebbly	o o o
sandy	. . .
silty	-.-.-
muddy	- - -
calcareous	⊥ ⊥ ⊥
dolomitic	⊥ ⊥ ⊥
cherty	▼ ▼ ▼
carboniferous	■ ■ ■
saliferous	# # #
gypsiferous	∧ ∧ ∧
tuffaceous	v v v
fossiliferous	ↄ ↄ ↄ

fossils

ↄ	fossils (undifferentiated)
Ⓑ	fossils - broken
Ⓖ	ammonoids
◁	belemnites
⛟	bivalves
▽	brachiopods
Y	bryozoan
⊕	coral - solitary
⊛	coral - compound
★	crinoids
⌂	echinoids
⬚	gastropods

sedimentary structures

erosional

∿	flutes
∇	grooves
↶	scour & fill
↶	load cast

depositional

≠	structureless (massive)
=	parallel bedding
≡	parallel lamination
≈	wavy bedding
≋	wavy lamination
///	inclined bedding/lam.
⌐	cross bedding/lam. (tab = tabular tr = trough)
∼	flaser lam./fading ripples
∿	convolute lamination
⌒	lenticular bedding/lam.
≻	wedge-like bedding/lam.
·.·.·	reverse grading
or ▽	(over thicker interval)
·.·.·	normal grading
or △	(over thicker interval)
⊘⊘⊘	imbricated

bed/layer contacts

———	sharp, planar
– – –	gradational, planar
∿∿	sharp, irregular
-.-.-	gradational, irregular
∿∿	disturbed

post-depositional

⌐	flame structures
⌐	sediment injection
◣	shale clasts
↶	load casts
⊙	pseudonodules
∿∿	convolute/contorted lam.
⊔⊔⊔	mud cracks (syn. = syneresis)
)|	water-escape pipes
⌣ ⌣	water-escape dishes
⊣	microfault
⌐	filled fracture
⊙	nodule/concretion

biogenic

⟨	bioturbation minor (0–30%)
⟨⟨	bioturbation moderate (30–60%)
⟨⟨⟨	bioturbation intense (>60%)
⟋	burrow traces
⅄	rootlets
⌢⌢	algal mound
⌇⌇	tracks & trails
⊤	borings
ↄ ↄ	fossil fragments

other symbols

( )	structure indistinct
(( ))	structure very indistinct
↑	interval over which structure occurs
⟨	disturbed section

selected geological symbols for mapping

25 ⌐ dip and strike of beds   ⊢ vertical   + horizontal

30 ▶ dip/strike of foliation, cleavage, schistosity

62 ◤ dip/strike of jointing   42 ⟋ lineation and plunge

⟋ observed   ⟋ inferred fault   ◄◄ thrust   ⟋⟋ strike slip

⟋ contact observed   ⟋ inferred   ⟋ uncertain

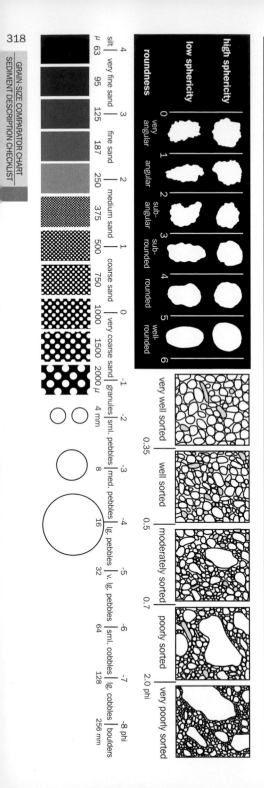

Sediment description checklist

Locality
- Record location details (map number, grid reference, name).
- Particular reasons for selection of this locality (e.g. type section, testing theory).

General relationships (field sketch)
- Bedding – way-up, dip and stike, stratigraphic relationships.
- Structural aspects – folds, faults, joints, cleavage, unconformities, intrusions, veins, etc (measure spacing, sense, orientation, etc).
- Weathering, vegetation – may reflect or obscure lithologies.
- Topography – may reflect lithologies, structures.

Lithofacies and units
- Note principal lithofacies present (sedimentary/other rock types).
- Note any larger facies associations, sequences, cycles, mapping units, etc.
- Note state of inclination and/or metamorphic grade.

Detailed observations (sketches or logs)
- Bedding – geometry, thickness, etc.
- Sedimentary structures (including paleocurrent data).
- Sediment texture and fabric.
- Sediment composition and colour.
- Any other information.

Questions and theories
- Note down any questions, problems, ideas, plans for sampling and lab work.

Bed thickness

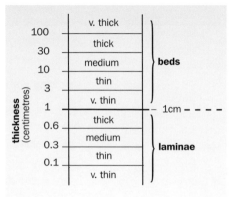

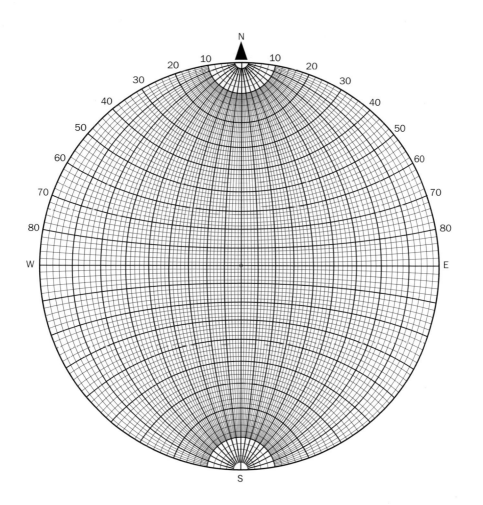

WULFF STEREONET
for the re-orientation of palcocurrent data

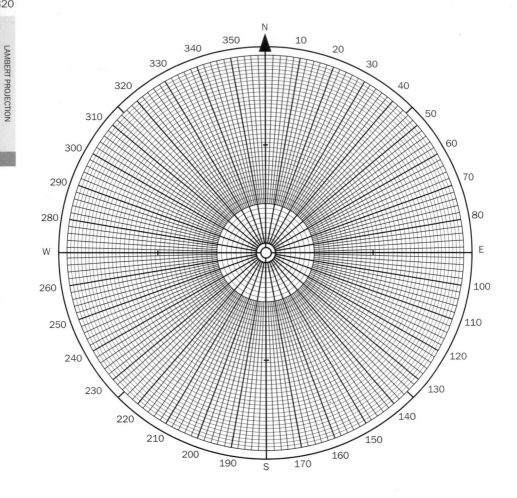

LAMBERT EQUAL-AREA PROJECTION
for plotting rose diagrams of directional data